A Woman Needs A Afterlife

Satty G. ©

Forward

I have written this book, firstly, to share a story about oogenesis. I am interested on the topics of oogenesis, and the menstrual cycle, because the process, and the cycle, both are a part of a woman's life and I think every now and then someone should say something about the two. I guess, for most of my life I have believed in the ideal woman, and so I also believe in the ideal oogenesis and if anything can be ideal about the menstrual cycle I believe in that too, -which arrives to the body of a young girl with a response from her that cannot be judged. I believe that on the topic of the body of a woman, and a young girl, not that much is known, because the body of a woman is ideal and complete and perfect, and so is untouchable by any idea. I was able to freely write about oogenesis, and simply dream that everything in the life of a woman and a young girl can be perfect for her, from her first day of life, until she has met a perfect pair for her and then on into the future. As for what else is in this book. For a while now I have been very interested in creating a chart of life, in our modern day, from looking at the world standing here on Canadian soil. In this book I do elaborate on what I think a chart of life should look like, and who I think should be in it, as best I can, and I do compare this new taxonomic chart that someone will make in the future here in Canada, to the charts of life we have created in the past, like the Scala Naturae. I think a Canadian woman belongs in a new chart of life,- with other species that are female,- alongside a Canadian man because no one has made a chart of life from the Canadian perspective before. Religion is also a theme in this book. What it means to be influenced by the western tradition such as the Christian faith, western art, and the mythologies we in the west often say we are inspired by to make art and to just simply believe in, like ancient Greek mythology and ancient Egyptian mythology, is a theme in this book. And lastly you will find in the chapters of this book topics like our night sky and outer space and writing on the seasons. I have imagined a origin story for why the moon revolves around the earth and why it begins its movement around our earth from its seat at deep east of our planet, and why that movement is on no orbit! In this book I have pursued the idea that if Venus and Mars do argue, and I do think that if everything was different about outer space, they would, then what would this competition between these two planets look like to the world, and so I have discovered that the myth about these two gods dueling has meaning to it. I have tried my best in this book to bring together many different examples, to conclude with the idea that a woman needs a afterlife: a woman surely needs ideas to become popular in the world about her body and her biology, at a modern time, in a modern world. A woman surely needs the female perspective on other topics about this world and universe, to become popular, always.

Thank you,

Sincerely Satty

Part 1

Chapter 1

The Menstrual Cycle

Even if a baby girl is born with 400 eggs in her womb, that could become children in her future, this story is about one or two or three to-be eggs that will surely become a child in the future of a young girl. One to-be egg at some point in a young girl's life will always be different than the rest, and probably two, and probably three. One to-be egg will be so different than the rest of the eggs in the womb of a young girl, that it could be a entirely different cell from the other eggs. And if the cell is not different than any of the other cells, -of those which begin to grow in the ovary of a young girl,- then for certain the oogenesis that it is a part of, is. Oogenesis may only happen in the life of a young girl, -or when she is a woman,- three times, or for how many times she gives birth. When that cell is ready to take part in oogenesis, no other cell exists in the ovary of a young girl. I do not know if it is better to be born with many eggs, or to be born with no eggs or to be born with one egg, inside your ovary, if you are a young girl. I do not know if it is better that one egg arrives to live inside the ovary of a young girl, and to take part in oogenesis, but I do believe that pregnancy is a lucky thing and it Does arrive rarely, to become a part of a mother's and father's life. The oogenesis that arrives as the life-cycle of the cell that does become a child, is unique.

Inside the body of every woman is a moon, like the moon we have in our skies. This moon is the primary oocyte, and this moon, existing inside the ovary, will be center to celestial events. Before it was a recognizable moon, -recognizable only because a celestial sphere full of stars circled it, shone on it, and gave it life,- it was a lone moon. And it could have had land on it and water on it, if that time in history did not pass when one moon came to circle the earth. It could have had land on it and water on it if there were going to be new earths, and if our earth, like a empty hotel room, was vacant for anyone to stay. I whole heartedly believe that our moon that has come to circle our earth has done this because of choice. Our moon, which we see in our night sky, still, is remaining in choice, to move around our earth. It does not move around our earth on a orbit, and so if a moon arrived inside the ovary by some wish or choice, that would be good. A long time ago medieval language made every celestial body a moon, besides from the sun. And so every woman wanted to give birth to a sun, so that her child would be as brave and unique as Apollo, the sun god. And when the fool from the major arcana, the one that performs in front of the divine court, the one that is the true troubadour, our artist, the antithesis, Dionysus, -the god that represents our moon,- increased in popularity over the ages, then every moon became unique. Today the moon may be more modern than the sun, and every woman wants to give birth to a moon. Today the moon may be more modern than the sun, because one moon came to circle our earth because our earth is appealing, industrious, magnetic, modern. There is one cell that exists as a precursor to the primary oocyte in the oogenetic tree, and it is the oogonium, and the world today has called the primordial germ cell it's predecessor, -but no one really knows if that existent is even a cell and even exists. Because I do not think genetic information is passed down from the mother and father of a child, to the child, in a overly literal sense, the oogonium, -which hardly can be seen- poses a question for me. And that is, is this precursor to the primary oocyte a unfertilized thought? And if so, is the primordial germ cell nothing? Would a unfertilized thought look like it contains genetic information? Or is peering into the oogonium, what it means to be looking at recessive genetic information? As for the primordial germ cell. That cell is surely round, but that cell is not old, and so I propose that that cell be called by another name, a more modern name. It could be called a moon. It could be called a celestial sphere. But because it also is new every time oogenesis takes place: that cell is never used, and it waits for nothing, you could call it a moon. And so that moon was there, inside the ovary, to pass down hollowness to a cell that is no animal cell, and no one did see this happen: *it was a celestial sphere fueled by imagination alone, it was the intangible, it was the un-handleable globe of the earth, it was a calendar to give birth to and to give birth.* If there is no primordial germ cell in the body of a woman, because the body of a woman is not animal, and the body of a woman is not old, because there is no beginning to the body of a woman, -even if certainly there is a beginning to the menstrual cycle in the body of a woman-, but because a woman may want to be a product of something modern, we will come to find that the phylogenetic tree which is oogenesis surely still has a one precursor, -in that chart,- to the primary oocyte, and we will come to find that we need this biological manifestation for a oogenesis that is ideal for every good woman, a oogenesis that has one major fork. Like a imaginary number there is a celestial sphere, which passes down hollowness to the primary oocyte. This celestial sphere is the perfect world above our constellations, above the most northern pole, above Queen Cassiopeia our cracker-jack. This celestial sphere is the untouchable globe of the world, a notion of a world that we can only imagine, when nature does not want you to know more. When no more information will come on the topic. And we can see, as the primary oocyte exists inside a growing Graafian follicle, inside the ovary, that hollowness has been passed down to the world of our to-be egg, and in no place in the universe does this happen, accept when we can imagine what our celestial sphere, our constellations will give to the moon in our night sky, a mobile of the moon and the stars we can still only imagine exists because our moon in our night sky still outshines the stars in our night sky.

    It is important to notice now that oogenesis, the process by which a child can possibly be born in the body of a woman, if a to-be egg is fertilized by a spermatozoon, is a phylogenetic tree. A phylogenetic tree is a chart which compares the growth, over time, of similar species on a visual platform, a chart usually used to locate the age of a species within the timeline of history. Gender is not a usual comparison on a phylogenetic tree. Usually, in a phylogenetic tree, gender is not a topic of concern. You will surely find in a biology textbook a tree like this, where the species that are compared, like for example dinosaurs, are simply suspended in time itself and there is a assumption that exists that every species on that chart is male. This is one of the reasons why those phylogenetic trees become boring over time. *Just imagine a phylogenetic tree that compares the origins of species and on that tree the animals are female.* I have always believed that if a man and woman were compared in one phylogenetic tree, as if we were trying to find some origin of both man and woman, as if we paired a existential origin of a man and a woman with the taxonomic origin of each, that the woman, on a chart like that, would have existed before the man, on a chart like that, much like how the homo erectus existed before the homo sapiens, or how the bird existed before some dinosaurs. I believe that a woman outlives and predates any man, because, it is a woman who gives birth to a man-child. Oogenesis is a phylogenetic tree that compares gender and only gender. On this tree, what is male and what is female has no face, because this tree Is sexual differentiation, and it does happen in the ovary of a woman. It is the smallest phylogenetic tree that exists, but it surely does portray a race that the two genders take part in as competitors, so that one will win and will be born out of the body of a woman.

    No woman could make do with a life of fulfillment to herself and the race of mankind without that one major fork in oogenesis. This one major fork is when the primary oocyte becomes a secondary oocyte, once pushed out of the ovary. I believe it is The major fork of oogenesis because I believe it is at that time that the gender of the child, who will grow in that cell, whenever he or she will be born, is made certain. If we were to idolize giving birth in our society and cherish it and make it important, discovering that sexual differentiation does happen in the ovary makes the life of every to-be egg something that we want to actualize. After discovering that sexual differentiation happens at a major fork in oogenesis, and so the life of a cell that can become fertilized is recognized for its potent life, we will all surely want to actualize the utter, fecund, possibility, -which is pregnancy-, every opportunity we get. That gender is decided at this fork surely makes us all want to believe that there are far less primary oocytes in the ovary of a woman and young girl. This major fork in oogenesis will not exist if we do not accept that there is a precursor to the primary oocyte in the biological process oogenesis.

    The primordial germ cell is what makes a woman animal, and at times there is something erroneous about that cell, because a woman is not animal. And this animal cell will remain, in waiting, inside the body of a woman? Maybe because the primordial germ cell, described in those words, is heavy. Or maybe it is because that cell, described in those words, is just too real and just too far away from religion. If there is no precursor to the primary oocyte, the resulting ovum from a oogenesis where something comes from nothing leaves us with a predictable pathway to a child, when there is something cosmic and heavenly about the birth of a child and the body of a woman that bore that child, and the body of a man that fathers that child. That heavenly sphere, who gives to the primary oocyte, exists a day away from the recessive gene that is the oogonium. The oogonium is a recessive gene because no celestial sphere did circle around it inside the ovary of a woman. When we are looking at the oogonium we are not looking at a moon with a shell, we are looking at a earth. When we are looking at the oogonium we are looking at something that the to-be egg could become, always, but never will. After knowing everything, the oogonium is a earth with land and water on it, but every woman in the world will finally decide that she would not want to give birth to it, because there can only be one earth, and we are living on it.

    However mysterious pregnancy, and bringing a child into the world can be, science has suggested that every woman should celebrate the birth of a child through a fecund statement, which is that the predictable pathway to the birth of a child, oogenesis, can be found in the biology textbook, when not long ago, everything that would happen to and unfold from the womb of a young girl was very private and taboo. The biology textbook was for everyone, and everyone shared in this fecund statement that though a woman is unique, is on pedestal, is not a man, that no woman is unique from another woman, because the physical reality, the biological reality of oogenesis, was readily available to all the world and to everyone who came to live here in Canada, in the biology textbook. This made a oogenesis that happens in the body of a woman very accessible, but it also made it easy for us to all perceive that oogenesis happens very quickly in the body of a woman, because oogenesis became convenient. This convenient oogenesis also made us believe that the process is exactly the same for every woman, and we will come to find that it is not.

There is a first time that the ovaries began to work inside the body of a young girl. And whatever that work was, it was nothing more than breathing, it was not ovulation. What if one ovary began to breath first in the body of a young girl, before the second ovary began to work. Like a solid gold or marble or stone idol of a woman came to life, as fast as possible and all at once a idyllic woman is born, out of clay, but, one ovary, fashioned first, begins to work first. The reproductive organs inside the body of a young girl are different than any of her other vital organs. The ovaries have a beginning, to work, at some time in a young girl's life, when she is no longer a baby. They have a life attached to the thinking, and reasoning and imagination of a young girl. And like clockwork, like a domino effect, after the first ovary began to work inside the body of a young girl, a muscular signal was sent throughout her body for the second ovary to begin to inhale and exhale as fast as possible. Blood was the communicator between the two ovaries. Before the menstrual cycle arrived in the body of a young girl her reproductive organs were still. Can you imagine that every month afterwards, only one ovary at a time takes part in the menstrual cycle: blood passes through the nerves that surround the ovary and the two oviducts and the uterus and is dispelled. The next month this happens to the other ovary and so there could be a *cycle of the second ovary*. We should make certain that even ovulation, - a term denoted to when the primary oocyte is pushed out of the ovary and into the oviduct-, could very well happen in response to the desires and fantasies a young girl has, towards her future, for a lover, or for having children. I believe the dreams that a young girl has can cause her own body to work. The young girl has a body that is inspired by fantasies and dreams that no diplomatic world can sublimate. I believe it is a very rational thought to think that in the body of a young girl, her one ovary will begin to inhale and exhale before the other, and that in itself is a powerful statement: the body of a woman does work like a well-oiled machine. Surely ovulation does not happen every month. Perhaps I just do not think that the body of a woman discards much, and for her to discard many to-be eggs that have not been fertilized every month? Do secondary oocytes wait inside the oviduct for a auspicious event to happen and it never does? During the period in the body of a young girl, blood is shed from all open pathways. Would a group of female cells, vulnerable because they are no longer being kept safe in the ovary, be pursued out of the body of that young girl by blood, maybe.

If a woman wants to make it into a chart of life while we know that at some point in her life she will house a phylogenetic tree in her womb,- if she is the ideal woman,- she will have to face adversity of all kinds. In the future she will have to compete, with the help of her Christian vows, against a world that might not believe that there is a precursor to the primary oocyte. Perhaps the precursor to the primary oocyte in the oogenetic tree is something rooted in the Christian religion. Do not the angelic choirs exist, echoing around a sphere?

The phylogenetic tree which is oogenesis Is unique for every woman, despite what the casual readily available biology textbook would make us believe. Oogenesis is like the backbone of any child, so long as every bone is there the length of the spine of the child is always unique. The oogenesis inside a body of a woman is unique for every woman, because the length of the phylogenetic tree is always different for each woman, something, if does not go unnoticed, can make a statement about the life of a young girl. When she met her first and only lover can be marked on just one oogenetic tree, and when she first fantasized about having a child: oogenesis not only maps the steps to the birth of a child, but also the events that took place in the life of a young girl, and the events that took place in the life of a woman, up until the child is born from her, because one primary oocyte may exist inside of the ovary of a young girl for a very long time, and so can the secondary oocyte exist in the oviduct for some time. A woman born in the western world will surely understand that oogenesis is unique for every woman because that is a woman who is not afraid of what modernity brought woman: for hundreds and thousands of years we have made art in the western world and it was pious, and the subject matter of art history was the body of a woman, so every woman born in the western world will certainly be prepared for what centuries of pious art on the topic of the body of a woman,- a nude, a muse, a mother, a lover, a wife, a sea nymph, a land nymph, a bather, a angel, a god,- will present to the world. But this year everyone shared in a fecund statement with us, when every woman born in Canada, wanted to believe that they were unique.

In the ovary of a woman there is a celestial sphere that exists before the primary oocyte is formed, as a precursor to the primary oocyte. The oogonium is surely the most fertile cell in oogenesis, but it passes down nothing to the primary oocyte, and so a heavenly sphere gives everything to the primary oocyte, because unlike the oogonium, the primary oocyte is cold and shelled. The oogonium may only be the image of a cell that the moon could not be: a sun with sun spots on it, or a earth with land and water on it. In the future of the to-be egg, we will never see the recessive information inside of the cell. Every woman should want to believe in a precursor to the primary oocyte because without that celestial sphere there would not be one fork in the center of oogenesis, which exists, on this tree, in between when the primary oocyte becomes a secondary oocyte. This fork exists during a period of time that the world has decreed is ovulation. This fork is a fork in the road, is a moment when a young girl takes a prophetic glance into the future, into becoming a woman. This fork is virility and independence for a young girl because, let it be known, that a young girl up until this very moment needed no involvement in her oogenetic tree by anyone, for that secondary oocyte to be born. A young girl can, if she so desires, make ovulation and every connotation that comes with the topic, disappear. And if someone was to ask me if the primordial germ cell alone, that celestial sphere, that imaginary number is our constellations circling no moon, I would have to say yes: it could be the historical calendar. That first celestial sphere, which the world has named the primordial germ cell, a wildcat named Tigerclaws says: *that is what a cell looks like that is not sexually fertilized by any thought process, or by any other life.* And because that cell is fertilized by nothing, it cannot be chaste, like the primary oocyte is, and it cannot be accomplished, like the secondary oocyte is, there is no fantasy and no idea to it, but it is feminine and it is celestial. And because it is inside the body of a woman, the following cell of oogenesis is feminine. The oogonium is the middle ground in a fine art painting. It is the Madonna and child in Giorgione's The Tempest painted in 1505, whereas the primary oocyte is always caught up with god, existing in a place like heaven high above the world, existing in a place like where the angels exist, wearing white raiment, high up on the top of a El Greco painting.

    The moon in our skies, in our world, is center to no celestial events, and is only the complacent existent who allows rather generously for the earth's shadow to pass over it when it does, and the sun's light to shine upon it when it does. The moon is not alone in a Lunar eclipse. The moon is not alone in any one event except, for us, the moon is a vital timekeeper, and when we think about this celestial body as portraying time itself, only then is this celestial body alone. If the moon was alone, he would be Dionysus, the dead god, the god if we can say, on the cross, the god who disappears unexpectedly only to return, who brings us frenzy and craze and mysticism and ecstasy. Many have called him Diana, not a dissimilar name. Why the moon was made a woman, sometimes, in mythology, was because in mythology the attempt was always to contrast masculinity and femininity and because we have always known the sun god to be Apollo, a man, Diana, as a woman, was perfect to assume the feminine role in a duality, because the sun and moon were always inside a region of debate: were they opposites or not. But for that one day when the moon is new, if the moon could be Diana when it is waxing, and Dionysus when it is waning, how are we to know, when the moon is not visible in the sky, who that new moon is? Because Dionysus is the god that leaves and comes back after a while, then every new moon should be that dead god, or could there be two different types of new moons when we know there's not? We have learned from mythology that there is always two contrasting figures that fulfill what it means to have opposites in this world, and we should be sure now that Is man and woman and that Is male and female. From within the mythological perspective, the moon was center to events. So, can we now, finally say, that the first events were Dionysian craze, even though these events happened during the day? In the modern day, could that be the shopping spirit that every woman enjoys? Is that shop 'til you drop? The moon Dionysus was center to a cult. One that in our oldest understanding of what is a tragic play was rival to that tragedy, it was a comedy. To gain the perspective of a other world view is why the to-be egg in a woman's ovaries is a moon. Our moon, the one in our skies, is a time keeper, it's a timepiece, it ticks away the time that which a sun will die, through the passage of the shadows across it's visible surface. The moon takes part in two plays. One as it's shadow increases to a new moon. Then the sun has died in a sentimental farce. But as the light from the sun grows big on it again and constantly towards the same direction does it go big and small, the continuity of this, that is a inevitable clockwork to the day that our sun dies, and it is no sentimental farce, but part of the second play for which the only playhouse is the moon. The moon itself deserves to be center to the celestial sphere. It is the moon amongst our stars because the moon accompanies our starry night but also because our starry night accompanies the moon: our moon deserves to be center to the same celestial sphere the earth is surrounded by and if we could envision how our stars in our night sky shine on the moon we are lucky. It is not that we have not mapped out every crater and flatland of the moon and made maps of the moon, because we have, but we don't have a globe of the moon, we have never created one for the moon, animated with our constellations,- like we have done with a globe of the earth,- but our moon is center to the constellations that fill our night sky too. Inside of a woman's body the moon can be center to a celestial sphere, and it is, and there the moon is center to a event, and there, there is no sun, and so there, the crust of the moon can be complacent to something else that shines light, and that is the stars of different magnitudes, that are blue, orange, red, white and yellow, the same stars that exist in our night sky: inside of the ovary the primary oocyte is surrounded by a celestial sphere which contains the constellations that circle our earth. If we say that the primary oocyte inside of a woman's ovaries is itself a moon, it is not because of any generic idea that there is a play in the woman's body, but if a man strongly believed that the woman that he loves has a body that is not made of stone, so that he can worship it, a man can make a mellow-drama of the body of his wife, and it would be true, just like a grandmother or mother of a woman can see a play in the body of her daughter. Inside of a woman's body, this ever present capability of giving birth, and its opposite, the bleeding in the womb of a woman, this all may add age to the body of a woman, and then may take it away: the menstrual cycle in a woman's body is a sure route to immortality for her, so some might believe a play exists in the body of a woman. Our reason for believing the primary oocyte is a moon is so that we can interpret non-somatic genetic information as celestial. Our reason for believing the primordial germ cell is a celestial sphere, is so that we can interpret non-somatic genetic information as celestial. The moon in our skies is center to a animated circus and we will make a handleable globe of the moon with our constellations painted on it, and when we do, oogenesis will not change.

    This calendar, this ever-present character like time, who presses forward continually, places himself everywhere. There is a calendar in every reasonable place, in whichever place that needs a deterrent from death, in whichever place that needs a opposition to death. *There may be a calendar in the top right-hand corner of the ceiling in my room*, -as adapted from a jazz song called <u>In the Morning</u> sung by Nina Simone in 1968- *there must be a calendar, also, in my womb.* There is no palpable gene passed down from mother and father to child when a child is born, unless it is a recessive gene. The recessive gene is like a whisper in the wind. It is a cultural trend. It is how the face of a Queen remains in circulation like a coin. The recessive gene is inherited from the mother and the father while the ancestors of those two people watch descendance take place: inside the ovary the primary oocyte is surrounded by a celestial sphere that moves around it like our night sky circles around us, and for that one time in the universe the stars of our night sky become ancestors to someone, and that someone is the child that Will one day be born from that cell. *Our night sky and the constellations that fill it up are a lot older inside of the ovary.*

    A white wolf says: *There are truly only two poles to a to-be egg. And surely like the earth has two poles, one is cold, but in the primary oocyte the other pole, the southern pole is hotter, unlike the earth, and this is why we are looking at the equivalent of a moon which begins its life cycle in the ovary of a woman, and not a earth.* It is very unique that the to-be egg cell has poles, as compared to all the other cells in the body of a woman. And it is very unique that the to-be egg cell has a equator, as if only the process of taking away from this moon, is preventing the cell from becoming a earth: this moon has a thick, hard, milky white shell and before water is found on it's surface some part of the entirety of the cell is taken away, as a first polar body. And before whatever is left of this cold shell harbors soil, and frost, and grass, and land, it is taken away as a second polar body. The divisive process which is oogenesis will benefit a world instead of a earth. The divisive process which is oogenesis will benefit a world where the moon can be center to a celestial sphere. The stars in our night sky are always trying to make a fossil of the moon, and they have already done that to the earth: the inherent quality of a star to shine light provides a future and reason for this behavior, which is a organism imposing on another organism. The stars will always shine light and will always try to impose the pattern they make in the night sky, on to whatever is below, as if their light can be felt by soil and grass. The moon in our night sky is always lucky that though it is center to the celestial sphere, -like our earth is,- no star pattern will cause it's crust to unravel because far distances away, there is one sun in our sky. Inside the ovary of a woman, and a young girl, there is no sun, and so the celestial sphere, which comes to circle the primary oocyte, will impose on the shell of that moon, and when it is revealed that inside of that shell is a cell, it is a Eureka moment! It is another moment in history when the stars receded. If the celestial sphere makes a mark on this to-be egg, we will find star clusters made a mark at the equator of this cell. In our night sky the stars are never divided. Lone stars there have always been known to stand out for being alone. Halves and pieces of entire constellations which randomly or by destiny itself did make their mark on the equator of this cell, of the to-be egg, are stars that will be remembered for being stars alone and they will be remembered by their names alone, -by the plasticity of this cell,- rather than being part of a constellation. These stars are surely weak because they have been pulled apart from a entire great myth, but at least in the life of the to-be egg they have a chance to finally stand alone. Some constellations are lucky enough to be housed in certain regions of the northern and southern hemisphere, and so in the life of the to-be egg it seems by good chance they will be remembered as a entire great myth, but then for that night, for when that auspicious event did happen to the cell, -which was when the primary oocyte became a secondary oocyte, was pushed out of the ovary- star brightness and sky cloudiness also impacts how bright the constellations will shine and impact our cell. And when children are born during the day, or if any auspicious events that do happen to the cell, happen during the day, there is always a night before or a night afterwards, and there is always a night existing outside of our plain view, for certain. What if the dominant genetic information inside a sex cell, however infallible it is, is not a reflection, a shadow more like it from the surface of the cell onto the center of the cell? That reflection would be of anything and would be of any colour if the primary oocyte existed in any other place in the universe. The oogonium contains genetic information for which we have no explanation as to where it came from, we can barely even see that cell in the ovary, but it contains information. We can assume that because nothing happens to the oogonium that nothing happens to whatever is inside that cell either and so whatever is inside of that cell could be recessive genetic information. The primary oocyte is the beginning, in the life-cycle of the to-be egg, of when a shadow will be cast by the stars that surround it, into the center of the cell, and that shadow is dominant genetic information. It is not unlikely that dominant genetic information inside of a plant cell is a image as well.

  Only one non-somatic cell begins as a hollow sphere with a tough shell, and that is a primary oocyte of a woman's ovaries. No cell in the body of a woman has a shell except for the primary oocyte. It is a rarity. It is something special. Perhaps no other cell of the body needs so much protection and safety as much as a to-be egg. And it is no surprise that this cell, which began with a hard rock-like shell will become the weakest cell in the body of a woman, when the shell separates, because it is virile, and frigid like a spinster. That cell will contain dominant genetic information that is a reflection of the cellular membrane of the cell, onto the inside of the cell. This reflection is a result of a particular pattern that is pinched on the surface of the skin of the cell. This pattern is more or less put onto the skin of the cell, when the crust of that moon separates due to the intensity of the stars that shine on it. The spherical crust breaks off at two separate times, first as the first polar body, and then lastly as the second polar body. The crust of the moon is cold, and so the light of the stars that circle it eventually cause the crust to become crisp and break off, because the crust of the moon -in the ovary, and when it is in the oviduct as a secondary oocyte-, can only hinder the rest of the process by which a baby is born, it must break off, because the spermatozoon must enter this cell. Our celestial sphere is filled with so many ancient stories, much more than I know, but I do know that there is only really a north and south to our celestial sphere. The primary oocyte is a uneven cell, where there is a feminine pole and a masculine pole, and the feminine pole is weaker and smaller, so the crust on that half of the cell will crumble first as a response to the heat of the stars. The stars that make mark on that half of the shell will for whatever lucky reason make a deeper mark on that half first and cause it to fall off first. For that lucky moment, only certain stars, star constellations, and star clusters will be the most influential. The feminine side of the to-be egg favors stories that are about infamous men, and infamous women. If masculine stories, those that favor the dominance of a male, in our night sky, are powerful at this time,- because that was necessary and that was important to everyone involved- like the belief in the story of Orion as a man hunter, then the to-be egg will be a boy because the masculine stories made their mark first. If we looked at the primary oocyte at this time, and we found Orion there, marked in some way on the feminine side of the to-be egg, - on the shell or the skin, the pattern will be the same on both places-, we will find that the nebulae in Orion's lower torso was not shining brightly and was barely visible at all during that night sky. If the stories that made their mark are virile, and so prominent female figures withstood the test of time with this to-be egg, -because that was important at this time-, and the belief in Orion as Lady Orion was a more powerful belief, then the baby will be a girl and so surely during that night sky the nebulae in Orion's torso was shining brightly. This is why sexual differentiation happens when the primary oocyte becomes a secondary oocyte, during what the world calls ovulation. And that is the orgasm that happens to the body of a woman, and that may not necessarily happen to the body of a woman with any interference or help of a lover. At this moment, the celestial sphere disappears, and the constellations and the stories that were weaker and less influential on the to-be egg are waiting to make their mark on the skin of the cell, when the second polar body separates, but they are now destined to make their mark. What it means for the stars that are waiting to make their mark on the to-be egg, is that they did not shine too intensely. Perhaps they were not red or orange or yellow, and they were rather blue. The second polar body separates from the skin of the secondary oocyte once the spermatozoon penetrates the cell. From the heat of the spermatozoon, the masculine side, the rough and strong side of the cell, is issued to separate from the cell, because it may not ever break off on its own. If there was no shell on the primary oocyte, a cell that has no phospholipid bi-layer could not retain its shape. Remember, the primary oocyte inside a woman's ovary can never be a earth, because in our world we cannot see the earth in plain view, we cannot make a play out of the life of a earth, and there has never been a realistic and rational and irrational and reasonable story about the origin of the earth. The stars are in the ovaries as a sort of blueprint, a security measure because the primary oocyte does not spin, so surely the world around that cell will. When the secondary oocyte does spin, when it enters the world of the oviduct, the feminine side of the to-be egg causes it to spin because it is lighter. Uneven balance in the to-be egg causes it to spin once it is pushed out of the ovary.

    Nothing comes or arrives before the primary oocyte in a woman's ovaries, but hollowness is passed down from a one precursor to the primary oocyte, as if that one cell, that precursor to the primary oocyte is friend to the primary oocyte, a comrade, a ancestor, who makes sure that, though we know we want nothing to be hollow, that in that small cell, there will one day be enough room to hold a baby. And this hollowness, this gift to the moon in the ovary of a woman, gives the to-be egg braces around the edges of the cell, like bones that support the body of a person, bones as thin as those in salmon or sardines, to ensure that when the cell does grow, that it will not break. When the moon in our night sky first arrived and began to circle the earth, it did so because it heard a voice of reason, a law to abide by, and that was the industry on earth. That was the hushing and shushing of turbines. That was the operation of machines that only stop if you put a end to them. Our primary oocyte is a moon, not a earth, and so the constellations would surround it as if it were a moon, like the one in our night sky. The succession of constellations circle the earth for the same reason the moon does, a response to sound, a response to industriousness and in the ovary of a woman, the celestial sphere circles the moon and resonates, and is charmed by this relationship due to the sounds the cell begins to make, like a crackling fire, as the thin bone-like edges of the sphere extend, and do not break. In the womb, inside a woman's ovaries the constellations circle the primary oocyte for the same reason, a response to sound, a response to a chance to be heard by the future, and because our constellations have a succession, Orion arrives first to circle the to-be egg, always. Like we don't know if the planets and moon and stars and sun are there when we don't shine light on them, the same goes for the moon in the womb, however, we do know when life events transpire for a woman, and from there, if we are lucky enough to know that woman, we can make assumptions about what is going on in her body. A woman has a life, activities, friends, love, a mate, there could be many reasons which could trigger the arrival of a moon in her womb. We have a older story for as to why the moon circles the earth, but because outer space is so vast, we can have a older story: the space in the ovary is small.

    The cycle prior to oogenesis is the menstrual cycle. The primary oocyte is a sex cell, it grows, like any other cell of the body, but the skin of this cell has no phospholipid bilayer. Just like you need a strong bicep to be a baseball player, a woman needs a strong sex cell, one which adheres to rules, and laws, to divinity, to culture, and this sex-cell does, by being equipped with a elasticity, but yet being the cell that would tremble the most, more than any other cell of the body, in response to invention. During the menstrual cycle, skin cells are shed from all open pathways, which includes the oviducts, but eggs that could be life in the future are never shed, but nor do the eggs nestle in the womb, many times the womb is empty: the corpus luteum is like a world collapsing on itself, but with no one in it. Because of the menstrual cycle, night time is protected in the womb, and though that bleeding ensures that the cracks and crevices of the body of a womb, can remain taught and dark, and soft and sensitive, that cycle can't prevent the womb from getting cold, *because you can still get pregnant on your period*, and when it gets cold, that is when oogenesis begins. When the stars start to spin in a woman's ovary, it is a celestial play. A moon forms from one cell that hasn't been discarded with the corpus luteum. This first cell, this moon arrives from out of the darkness; that primary oocyte has a shell that is cold because of the darkness that surrounds it.

    The stars of our constellations circle this moon, and as they shine on this cold shell, they incise, with the brightness of a star, onto the crust of this primary oocyte. This primary oocyte, this moon, now is engraved, now is carved, it is a carving, and the lines become more pronounced, deeper and deeper-even though it is incremental,- until a auspicious moment arrives in the life of this woman to make a lasting mark. Tracing a map onto the shell, over and over, a pattern, like a blueprint, the stars mark on the to-be egg. When a auspicious moment arrives, one part of the shell separates from the primary oocyte leaving behind a pattern from either north or south of the celestial sphere. The markings may leave a photographic image on the skin of the cell. *The shell of the primary oocyte is being molded and is made out of clay,* and the result is a image on the skin of the cell that can hardly be seen, not unlike how a fallen leaf in fall can leave a stain on a rock after it rains, something being just short of becoming a fossil. The image on the skin of the cell was received by the heat of the cell, from the stars, of the very star constellations that were the most powerful and influential at that moment, as the shell parts away, and the result is a shadow inside the cell as what is inside the cell responds to the image on the surface of the cell. With almost nothing left to be seen on the skin of the cell, the genetic information inside the cell, is just a shadow and a memory, inside the cell, of the events that transpired inside the ovary. The skin of the cell is warm and the shell is cold.

As the stars incise onto the shell of the to-be egg, the shell weakens from the pressure until it breaks away, first one half and then the rest: the first polar body and then the second polar body. It must be certain that both polar bodies do not leave without benefitting the egg. The first auspicious moment to arrive to a woman's womb is caused by a first sexual initiation but a woman can carry a primary oocyte for a long time, even until successful sexual intercourse, and these two events, of the to-be egg to become a secondary oocyte, and of that secondary oocyte to become a ovum, can align and happen quite succinctly, like a domino effect. As for that first auspicious moment, if it does take place at a time prior to the day sexual fertilization did take place in the oviduct, like a child that thinks she is pregnant before she has ever had sexual intercourse, a sexual initiation causes the possibility of a woman to become pregnant on another auspicious day. Like a child that thinks she is pregnant before she has ever had sexual intercourse, the first auspicious moment, when the first polar body is formed is when we all discover that the woman is pregnant for the first time. This auspicious moment is the moment you find out you are pregnant. The half of the to-be egg, where the half of the spherical cell reveals it's skin first, is the half of the egg that is weaker, and it certainly is where the spermatozoa will tunnel through, to impregnate the woman. The most virile and masculine stories that exist from our constellations leave a mark on the to-be egg, and in response to them, the child will either be a girl or a boy but this is decided when the first polar body is discarded because the first polar body does not leave without benefitting the child to-be born, this certainly is sexual differentiation. The world out there is full of diversity, but that child should be born with one gender, the most virile parts of our constellations will ensure that if no one else can. We only have two genders: how long would a woman have to wait for the decision to be made, which happens inside her own body, as to which sex will be born. A woman is not indifferent towards what is happening in her body. A sexual differentiation which is said to happen in the embryonic sac over exhausts stamina. Remember, there is a north and south to this map of the moon, there is a northern and southern hemisphere to this map of the moon.

    When the constellations first arrived to circle the to-be egg, it would seem as though that was a event of a lifetime, but if a first sexual initiation coincided with this event, it is what is happening outside of the body that seems more exciting. When a husband and wife try and try again to cause a ovum to be formed in the body of the woman, so does the woman try and try when she is by herself, she tries at everything. She tries to be right and not wrong. She tried to follow the rules her family and ancestors have issued for her to live by in the modern world. Surely all these events can be a first sexual initiation, and sometimes there is only one first sexual initiation which results in pregnancy, and sometimes that is not the case. Once the primary oocyte adopts a photographic image, and is latent, waiting, -and turning as a response to the dizzying speed of the constellations which did turn around the cell before,- as a secondary oocyte existing outside of the ovary, after it has been pushed out during ovulation, surely we can see that cell with a ultrasound, *and then every woman will be charting her own sexual experience.* Once the moon adopts a full shadow of genetic material, inside the cell, during the very beginning stages of pregnancy, after sexual fertilization has occurred, then if you are in fact that pregnant woman, you will be able to see information about that growing child through a ultrasound.

Other celestial bodies in our solar system and for as far as the eye can see, are just that: celestial bodies. We only have two celestial spheres in our vicinity, and another one but that is a man-made globe and it is not in our vicinity: when we are standing on the soil of the earth, we can only see so far, so far into the fields so far into the night sky, but the man-made globe can not suffice for one gaping awe into the landscape. We cannot make a address to the entirety of one gaze into the universe when looking at the man-made globe, for we have to spin it.

We have the celestial sphere that houses our constellations, and we have the moon itself. Earth is a celestial sphere too, but we cannot easily view it and if the constellations are trying, are trying to make a photographic image of the one unchanging pattern on to the earth, which they are, which is the succession of constellations, the moon is often too bright for that to happen. *A place where a image is produced where one organism imposes on another*: there are not often many places that happens in this world, and with one moon, our earth will never become a fossil, and with a thin milky skin, and a tundra-like shell, our primary oocyte also never becomes a fossil underneath the heat and the pressure of the stars, because of course, the shell, burdened under the light of the stars, breaks away, and that is when our moon in the womb becomes so bright that it outshines the constellations that circle it. The moon and the earth, both have some sort of crust and layers of the variations of it underneath, our celestial sphere is hollow, like the inside of the primary oocyte. The moon has two mythological names, Diana and Dionysus. Because the moon has two mythological names, because both these names are for one man, for one of the names, Diana, we've crafted stories as if we have envisioned the feminine side of a god. Because we have stories that are of both the masculine and feminine sides of this god, the moon can remind us of a celestial sphere that has a northern and southern hemisphere, where one is feminine and the other is masculine. If for the primary oocyte, there are two poles, the northern and the southern pole, and the southern pole is warmer than the northern pole, then the feminine pole, out of the two will be the northern pole because the crust of that moon has to become cold enough to break off, firstly, and from being cold, respond to the heat of the stars that encircles it, and so the colder pole will break off first. If from the very beginning, of the life of the primary oocyte, the southern pole is warm, and stays warmer than any other part of that to-be egg throughout its life-span, then it surely will need interference from a outside world to break off, eventually, otherwise it will remain stuck to the southern pole. I have always thought that the northern hemisphere, of the celestial sphere that surrounds our earth, is feminine. Maybe because whenever I think about what is most north in our world, I remember a ice queen, or Queen Cassiopeia. If the celestial sphere has a feminine side, it just simply means many stories that makeup the constellations in this sphere, are stories about women masquerading as men, because during that time in history those figures were prominent, but to the primary oocyte, there truly is a weaker side to this cell; the feminine side, the side that breaks away sooner and maybe even much sooner. This quivering cell is everything, is powerful, is bendable but not breakable, but it is only virile whilst having a shell. The most virile women in our history make a mark on this shell, while shining through our constellations, but the most virile parts fall off first. In honesty, I have not seen a part of the night sky that is very feminine, perhaps the Pleiades are a feminine star cluster, but for as far as femininity goes in my perspective I think lady Orion is virile and feminine.

The skin of the primary oocyte is always feminine, and the same goes for the skin of the cell of the secondary oocyte, and the ovum, and the zygote. Even the embryonic sac is feminine.

Remember: we did not make a globe of the moon. Who can compete with the most best Image of the moon, where when you envision the moon in your mind, afterwards, after you saw the painting of it, after you saw it in the night sky, could you then remember it's gender? Every living thing has a gender, either male or female. If we didn't have so many stories about the moon, myths about the moon, perhaps we would not know what gender it is. We have many stories about infamous trees, like the cypresses of Van Gogh's paintings. Do we care to find out what gender those trees are?

There truly is a masculine and feminine aspect within a man and within a woman, there truly is a masculine and feminine aspect within the bodies of both. It is a shallow misconception to think that there is not. There Is a shallow misconception which is true, on the womb of a woman, and on the belly of a man, which is that both are gendered. The belly of a man is masculine and the womb of a woman is feminine! When a woman is on her period, there is a shallow misconception there, because she does not discard a to-be egg during it. Inside of ourselves, we are not divided by any gender, but there is a shallow misconception that has a gender in the womb of a woman, which is that every yes and no, every yea and nay about the body of a woman has no accurate explanation to the world, and to the on-looker, when that explanation comes from the mouth of the woman. During oogenesis, the feminine side of a to-be egg breaks away first when the to-be egg is a primary oocyte, and it is a very shallow misconception that, that part of the too-be egg is not virile. And surely, if a woman has a womb full of pickpockets, like what Paradiso looks like on the cover of Dante's <u>The Divine Comedy</u>, that is a shallow misconception but may explain why we believe that a woman has four hundred follicles- which are to-be eggs- in her ovary, when she is born as a young girl: *because the ovary looks like it does*. If the wrong egg is made to be a to-be egg, by accident, by mistake, that egg may just be eaten up by the corpus luteum, and so our idea of a timely ovulation, -something we should want to believe in if we planned for a pregnancy, something that is a disastrous idea if the wrong egg was made to be the to-be egg,- will add a unreconcilable age to that womb, will add one pocket to the womb in the place where the egg will not hatch, in the uterus, and our timely ovulation will add one pocket on the outside of the ovary where the secondary oocyte is to stick but won't because of our successful corpus luteum: this is the example of a ovulation that takes place inside the ovary of a woman with no result, with no rhyme or reason. Besides, the face of the ovary, where the ovary faces the oviduct, the outside skin of the ovary is already bumpy. If a shallow misconception was located anywhere in the body of a man, it would be in his stomach, it would be that the way to a man's heart is not in his stomach, but the skin on the lining of his stomach is smooth. The period before sexual fertilization can cause the removal of the first polar body, because if that spermatozoon could win, it would. Like a race every man should want to attain and win over every one of a woman's, a young girl's, fantasizes. The second polar body, -upon sexual fertilization,- breaks apart under the bright light of the stars, like the first half did,- but these stars are blue and of the smallest magnitude- and is released through blood spotting during a successful pregnancy. There is a possibility for the celestial sphere that surrounds the to-be egg to make a mistake, and make mark on a degenerate egg. Hopefully, for a woman, she will not attain many pickpockets, no woman wants a womb full of pickpockets, and it is a shallow misconception that the body of a young girl, that the body of a woman, and specifically her reproductive imperative and the organs that are involved, do not take a prophetic glance into the future.

The period just before pregnancy begins is not part of any menstrual cycle, it is part of oogenesis because it is that period that did not do away with the primary oocyte, and it is that period which did not do away with the secondary oocyte. It was that one period that was on the chart which measures hormone fluctuations before and during ovulation that you can find in a biology textbook. When the womb is bright, from the shining of the stars that are following in succession around the to-be egg in the ovary, -and there is a zodiac there, but no woman cares to know, just then, about the birthdate of the baby, because it is a secret and everything else going on in the life and the body of a woman is so exciting,- the cold heat of the stars causes the shell of the cell, the shell of the to-be egg to become crisp, because it is already cold, and do not forget, the skin of the cell inside is warm too. It would be easy to say that the two polar bodies formed, never go away, because a process so complex, appears as if it could only happen once, and because the polar bodies are not discarded routinely.

As for every woman and every woman's capability of giving birth, the menstrual cycle and oogenesis is the same in every woman, the only difference is, is when the process starts from the birth of a baby girl and from the birthdate of the mother of the child, and from the birthdate of her mother: this unique menstrual cycle, and this unique oogenesis, and the arrival of these two processes in the body of a young girl is inherited, a congratulations to a long list of mothers. What makes oogenesis and the menstrual cycle different for every woman, is when that first cell began to grow in the ovary of a young girl and when every auspicious event happens for that girl afterwards.

The archetypal image of a mother can never be discarded with a degenerate egg, that is the religion that existed a priori to Christianity, the religion of the woman.

When we inherit our personal physiological characteristics, character traits from our parents, and when we inherit our personalities from our ancestors, and when we first hear their music and their charm and their ideals, then we are chained to the idea of mankind, and because surely we hope that women can till the soil, then a woman is no longer a separate species to mankind. Does anyone have a suggestion to who else should till the soil; surely our children, and not every young mother knows how to farm yet.

When the child is beginning to grow, in the womb of a woman, this woman is a archetypal mother, meaning she may not need to flee from her eternal child-like tendencies to ensure that a child will be born in control of his or her world, and it definitely lies with the sincere immaturity of a archetypal mother to want to name a boy child with a girl's name too. Surely the mother of Diana, our Dionysus, has a eternal child, and I do think that the waxing of the moon is the feminine side of a god, and the waning of the moon is the masculine side of a god, because when the moon is new, it is Dionysus who has left us. Diana is the name of Dionysus during youth, and so like throughout mythology we have seen the life of a god, Apollo, from childhood until adulthood, we can also see the life of Dionysus. We except we have never seen the feminine side of Apollo, unless you Do know that the feminine side of Apollo is a sweet cherub. So, then we Have seen the feminine side of Apollo.

Many people want to give birth to stars and constellations in our night sky, but our stars only really want to compete with the sun-shining of the moon at night. When the to-be egg receives the sperm, that bright shining cell moves to where it will grow, latches on to one side of the uterus, but it is still bright in the ovaries for the entire pregnancy until the woman's period returns after pregnancy, to bring back darkness to the womb, to attach the womb to a natural and consecutive rhythm. During pregnancy, is it too bright then to see the celestial sphere turning in the ovary and is that why the celestial sphere disappears? Because only one ovary is involved in each pregnancy, it seems even more obvious that one ovary lies dormant during every period. The phylogenetic tree which is oogenesis only exists in between the primary oocyte, -inside of the ovary- and where it will go, to become a secondary oocyte, in the oviduct. *It is a chart that physically exists across the skin of the ovary, between the primary oocyte and the secondary oocyte, and when the secondary oocyte is fertilized, it disappears. It is a bone structure so thin and rigid* that only one man can make it disappear, and that is the man in love with a woman, during her singular lifespan. And if one more man can make it disappear, that will only be her boy, if the child is born a boy. One ovary lies dormant during every period in the vision of servitude to a greater good. If one ovary out of the two in the body of a woman is a back-up, like one kidney or one lung out of the two can save the life of a person if the other cannot work anymore, then so be it, but if the reproductive organs in the body of a woman represent something symbolic, then maybe there is another reason for a woman and a young girl to have two ovaries in her body. If form follows function throughout the biological processes of the body of a woman and a young girl, is there a need for two ovaries to take part in oogenesis at the same time? The reproductive organs in the body of a woman are different from the rest of the organs of the body. Perhaps there are two ovaries in the body of a woman so that one could represent night and the other could represent the day. Perhaps these two organs are in form and function more symbolic, especially when we know that night and day are environments that anything which needs to grow, does need to notice: *somewhere in the body of a woman a seed is fertilized*. Both ovaries together represent a night and a day on a twenty-four hour clock, but as for which ovary is the night and which is the day is a riddle because in the bible it is stated that god said 'and the evening and the morning were the first day'. Maybe one ovary is the dusk, and night, and the other is the hours that lead up to dawn, and even some hours afterwards which are the morning, until 12 pm on the face of the clock, and so the afternoon is never accounted for by the work of any ovary. But, still, it is certain that one ovary began to work first. And that did not happen to the lungs, and that did not happen to the kidneys when they first began to work because they began to work in unison. A wildcat named Tigerclaws says: *and so does happen in the body of every female species.* In no other organ inside the female body, does the cycle of a cell exist. Not in the lungs and the kidneys. Inside the ovaries a cell has a lifespan like no other cell in the body, and that lifespan is a cycle, this is why the two very similar organs, the two ovaries, have to begin to work at a variance in time, otherwise these real and physical cycles, which is oogenesis, would cause the body to be weighed more on one side. Does the body of a woman need both ovaries to take part in the menstrual cycle each month? Surely it is better that one ovary represents night.

    There is a first time that the moon is formed in a woman's body, I think it can happen before the woman gets her period for the first time, because it is a innocent process. Pregnancy may not be innocent, if a woman doesn't want it to be, but the formation of a moon in the ovary of a woman, in the ovary of a young girl, is innocent. But women are not born with a egg waiting to be born. Which young girl would like to carry the burden of knowing when she will give birth? Even though it is a luxury to be able to know this if you are a woman or young girl, -for every time you will give birth-, especially in the face of the modern world, especially in the face of modern day medicine and the science of the modern day man, but it is also important to recognize that the life of a woman, who was once a young girl, is always accompanied by a realization that her body is always in a optimal state of possibility, that which concludes with a responsibility: a young girl must always know that she can very easily get pregnant and will need to take care of the child, and that will be enjoyable, but after that, is she no longer a young girl, with a sense of freedom from adorning no cycle in her body? I think it is a big deal for a young girl to become a woman, and that young girl that you were, houses in her mind something close to a plunder: if you want to know anything about style and culture, and what is important and special and how to act to be politically correct, ask her. And if you were to ask her if the religion of womanhood is paired with Christianity, surely she would say so, and planned pregnancies are a long-held belief in the religion of womanhood.

    If a man, the husband of the wife, has a hunch as to when his wife will give birth, way into the future, the pedigree of decreasing chance, inside the womb of a woman, did in fact benefit the man. It has almost become, that sexual fertilization itself, is a common language for all that exist, where we all understand that a child can not be born without the contribution of a father. The second fork in this pedigree of decreasing chance which is oogenesis ensures that the secondary oocyte will be vulnerable, sticking to the outside of the ovary, and when everyone knows that young girl, who houses this entire tree, and that she is at this time in her life, when the secondary oocyte is available, formed and happy, everyone, every family member will be on the edge of their seats to accomplish a goal for this go-getter type of woman, now that they can see that sexual fantasies and desires of this young girl are coming alive and have come alive. And if this type of woman is a introvert, then the feelings of her family are still the same: everyone in that family wants to accomplish a goal, of the inevitable for that young girl, which is pregnancy once that secondary oocyte is available, everyone in the family of that young girl feels the need for this opportunity to be met by the lover of a young girl and this is a planned pregnancy. This second fork, the smaller fork in oogenesis, this part of the process of degeneration in the to-be egg, because it requires penetration, ejaculation, for the to-be egg to become a ovum, this part of pregnancy is like a safety net for a man that marries a virile woman, because if events transpired any other way, a woman that strong, and that head-strong would act like she doesn't need a man. A man and a woman have to share embraces for this pregnancy, but a woman can raise a child in a house full of women, excluding the father of a child, if she wanted to, if there were too many arguments between the father and the mother of the child, a woman will run to the cult of woman to find sanctuary for that child. A second sexual fantasy, one that is different for the woman, is different for the young girl, is guaranteed for the man she loves and that is to fulfill the desires of a vulnerable secondary oocyte. What woman would like to carry the burden of knowing when she will give birth? Maybe that is something, a burden that her mother might want to carry for her daughter, or a grandmother would like to carry for her granddaughter. If a woman would like to adhere to a virginal calendar, it is very optimistic and opportunistic for a woman to believe that sperm die quite quickly upon release in a woman's body, and also, that the primary oocyte only comes to be at a certain time during the younger stages of a woman's life.

    The first sexual initiation for a woman could be masturbation, which some women do not come across to understand until later in life, if ever, but a fantasy is even younger. A young girl can fantasize about anything that is good for her. She may even fantasize about being successful in the world and that is irresistible and those are real sexual feelings that cause heat in the sexual body of a young female. If a woman never masturbates, say in her entire life and specifically before penetration, and then takes part in sexual intercourse with her lover and a ovum becomes the product, perhaps the woman did not want to decide just what sex, out of the two sexes, will be born out of her body at that time,- and this is why there is a cult of women who will try their best to encourage growth in a young girl, not necessarily sexually or in relation to the body, of this young girl, alone, but in relation to her mind, so that she will not be raped by the man she loves so that she will conquer more often if always, in bed, when it comes to fulfilling her fantasies,- because children are often born in a response to the fantasies that a woman or man apply to their own ideas of a future for their children. Sexual differentiation will still happen as planned, in the ovary of a woman, even if a woman did not want to decide the sex of the child into the future; the feminine side of the to-be egg will still break away first; the most virile, prominent, constellations of that night will make their mark, but they will be in the guises of male role models instead of female role models, and that is a beautiful thing too. Orion will be the man with the club in his hand, instead of a woman with a nebulea in her womb but remember that Orion is and has always been a woman. If it is totally random as to which sex will be born and when, from the womb of a woman, then there is no such thing as destiny, and just because you may not want to know the sex of your child before the child is born, it does not mean that the arrival of the child and the sex of the child was not destined to be what it is, as decided by the night sky. And as for if the rape of a woman, by the lover of a woman, can cause the birth of a boy, instead of a girl, that is kind of a old mother's tale. And as for what rape is between two lovers, if a idea like that would ever come up in the world, where we can imagine this happening between a man and a woman who are lovers, in that situation the woman was not certain that at that time or in that moment she did want to have sex with her lover, and she was taken, and she was harnessed, by him, and she was held down. At that time, she was nieve for whatever reason. And she was taken and she was harnessed by him, and she was held down, at a moment in the western world when a woman just did not want to give anything or did not know what to give or just simply decided to relinquish control. The lover of a woman, in the western world, will never rape that woman for the love of a god, and for the love of a superstition, or in offering to nature itself. The relationship between a man and woman in the western world is private, the lover of a woman may subdue the woman, may rape the love of his life, but only for himself, so that he, alone, must take responsibility for his actions, not god, not nature, not anybody else, so that that act won't be a crime. And if a woman did not plan her pregnancy, was not financially prepared, was not planning to fit a child into her life and schedule during that time, and she did get pregnant, and she did not spend the last two years planning for it, well that is rape. Here in the western world a man and a woman will still fight about what they believe is wrong or right, and they will do that in the bedroom, in their sexual relationship between each other, and sometimes a woman will lose, but everybody loses some time.

Uro, a brown bull, shows up in Gaugin's painting <u>The Vision After the Sermon</u> painted in 1888 as a unmarked man. She may be a unmarked man because she, existing in two phylogenetic trees,- in one that is taxonomic and in one that outlines her career throughout the database of art history, which is a phylogeny because this career describes her evolution as a prominent artistic figure throughout art history and begins with the original cave painting of her which is in Lascaux- was able to race her mate, and give birth to the sex of her choice first. She may be a unmarked man because she showed up in <u>The Vision After the Sermon</u> before her mate could. Her arrival in this painting proclaims that a man may be put in detention to usher the arrival of her dominance in the night sky, in life itself, and that man is Taurus, and this can happen for any woman. A unmarked man is often a woman who could plan her pregnancy, and her entire lifelong career.

Oogenesis, though is equivalent to the definition of sexual differentiation- as a phylogenetic tree-, should not be sublimated to be defined by that part of the process alone, which is what defines the process which is as real as being physical: sexual differentiation. It would be rape to define oogenesis by one outstanding quality of the nature of the bone. It would be rape to define oogenesis as sexual differentiation alone. But the sex is assigned while the to-be egg is in the ovary: oogenesis is the story of the sexual and fantasmic life of a woman before her baby is born.

Oogenesis contains a fork in the road, a wishbone even, something that our worldly culture as a whole does decide about. When we leave it to a man alone, to coordinate our pregnancies, as women, we must leave ourselves vulnerable, palpable, to injury, to death to ensure that our girlchild can be wrestled down to the ground by a angel, and not die, and not grow up to be the marked man, not grow up to be wrestled by anyone else, not grow up to not have a wish come true, which is to give birth to the sex of her choice. We promise you Uro the bull will show up at the time of relieving a marked man who was born a boy- a boy born to be a king- and not a girl: this is the olive branch that the most infamous female in the world gives to a man.

Each time oogenesis crosses the path of a woman, she inherits a phylogenetic tree, something the to-be born child does not inherit, something the man in love does not inherit.

If we make the excuse that a young girl gave birth to a young girl, not everyone in the world will protect that excuse. Now, this is a woman who will defend her pregnancy. Not everyone in the world will protect that pregnancy but most definitely in the western world we did because each and every woman born in the western world, born to give birth to a young girl has a eternal child and that eternal child will make sure that this young girl can decide the sex of the child before pregnancy, if she wants to, even though she does not know and has no proof during the time that her womb is growing, of which sex will be born first, neither during her early years when her womb is growing, and neither during the time that her womb is growing and will not stop growing until her child is born. This is important because some women choose not to discover the sex of a child before the child is born. Perhaps it is only a wish that a woman makes which causes sexual differentiation prior to conception.

Uro shows up in <u>The Vision After the Sermon</u> by Gaugin to keep her crappy phylogenetic tree alive, one that begins with her cave drawing. At least the evidence to prove that she has a taxonomic identity is in that cave drawing, and this is when a phylogenetic tree has a past. Uro exists throughout art history, but she has a zoological history as well. Her competition is with a brown male bull Taurus. Her phylogenetic tree is crappy because it has never been calibrated and it shouldn't be, no phylogenetic tree should be calibrated, but surprisingly oogenesis could be and should be, perhaps because it is the smallest phylogenetic tree in the world. We can bank on Uro and decide to protect a crappy phylogenetic tree, which a long time ago was oogenesis, and would be oogenesis if the process remains taboo. It is up to a woman and her family and her lover to calibrate it or not.

There is a calendar in every reasonable place, and every place where it is needed. The to-be egg is surrounded by the constellations, but the first polar body is separated from the primary oocyte due to circa, one day, one moment in the life of a woman, one day, one moment in the life of a young girl: the moon in the woman's womb comes to gain a definite sex due to the traditions of the time, maybe even before any penetration. That is something we can't take away from a child that will be born, that is how a child is born modern. Our present day makes a mark on that child. If we can say that the modern-day woman is more virile, then the child born whether the child is male or female will be defined by that. But the modern-day woman is not more virile than the women we have learned about from old stories we have come to celebrate, from our myths, that outline our stars in the sky. It would seem as though the 365 day calendar is enemy to a woman, but that calendar is both enemy and ally to a woman. When we are racing against the clock in the modern day, we can celebrate holidays that everyone celebrates, if we so choose to, because sometimes the people that we love celebrate those holidays! Did women not work hard to shop, did women not work hard to admire the people that they found to have good qualities, like Jesus made the apostles fishermen of men? The 365 day calendar also houses the seven day week. And this week is what made a woman weak, because months, in this calendar, have been divided despite ironic occurrences. And the menstrual cycle is ironic, because convenience is a desirable accomplishment for both woman and man, and the monthly arrival of the period is not always convenient for a woman, and the most convenient invention we have ever made is the menstrual pad, but a woman may still not want to work on her period. But those pads are made out of cotton, that is a genius invention! The constellations are old and their stories do make their mark but we can't hide away from the 365 day calendar because a woman is made powerful and vulnerable according to that calendar.

The ovum results with a full imprint of genetic material on it's cellular membrane and a shadow of this information is inside the ovum: every victorious star makes a shadow inside, when once it was light that they shone, now they cast a shadow. After the second polar body separates from the to-be egg, a full shadow is cast, from the mark the constellations made on the cellular membrane, -through the shell-, from each dominant star and constellation, *on to the cellular membrane of the cell*. We are accustomed to imagining our stars as only shining light, but in a place that becomes brighter than them, they will cast shadow. In the ovary, only sexual differentiation waits a day in a year to happen, housed at the major fork of oogenesis when a primary oocyte becomes a secondary oocyte, but in spermatogenesis the entire process waits a day in a year to happen. The production of sperm responds to every single day, every day a man works, thinks, lives, breathes: there is a calendar in every reasonable place, and there is a calendar in the life of a man and his body, I just do not know anything about it. What happened in oogenesis, something that was invoked by a spirit world, by a world full of mythology, a statement made by the stars alone, all disappears: oogenesis ends when a secondary oocyte becomes a ovum. That moment in time changes, like a systematic law was forced upon the world, and let go, -when a secondary oocyte becomes a ovum upon fertilization,- so that the egg can take part in the rest of it's life to face the odds and to face chance, so that a baby that is born can encounter trial, whether that baby is born royal, that baby will encounter the world and will still walk the many various paths to find peace with the generic calendar. If there is a question as to who, out of the two sexes of our kingdom, causes us to question just what it is that is a species and what it means to be a species,-is it the man or woman?- when we look at sexual fertilization, for the first time ever, we may find the spermatozoon much more exotic, and for the first time ever, even though the lot in life for a woman is riddled with many questions as to why her reproductive capabilities resound in almost everything she does every day, when we look at the spermatozoon we may find that this part of what will become the life of a child, is more animal. Throughout history we have seen man, and his naked body, and we have seen woman, and her naked body, and both were ideal. But the stories that came about, about a woman, always described her as something different, and unique. Surely it was because of the reality and situation of her body, which is that she lives forever in the constant capability of being able to give birth, and her womb will house a child, and she lives every month in between a period. But through the infallible microscopic lens we see a aspect of pregnancy which is the spermatozoon and that is a exotic cell, especially after we know that the to-be egg is hollow first, and that may not be the same for the spermatozoon. However infallible sperm cells surely are, many times a woman will not get pregnant at the first try at sexual intercourse, but once a young girl has come of age, she is always ready for a possible pregnancy, surely bodily, and what is a must for a young girl is to be emotionally and mentally and culturally prepared, and perhaps that is why by fateful chance alone she has to spend more time worrying about all of the responsibilities that sexual encounter may cause her to endure, on a daily, so that the duality of the two sexes remains, because a man will shed sperm cells daily, and he will do that without a woman. A man has unending sexual rituals that a woman can only dream of uncovering. All of the sexual fantasizes that a man has, he can cause to fruition, once he ejaculates, whether they are complete successes or incomplete thoughts, but a man can cause those feelings to leave the body with such a finite result that he can make a comedy out of such a routine, or he can make any story that is handleable, feelable and touchable, from his sexual expression, a comedy which is not easy for a woman to be a part of because she will always remember that her body may go through a nine month responsibility where she can even die giving birth, if she has sex. A man can issue his real life concerns quite quickly if he is virile, but a young girl, and a woman, will always have to balance the obsession with two lovers, one a man and one a woman, and that takes a longer period of time throughout her lifespan to accomplish and it is something that she will always need to accomplish. Now that we know that oogenesis is very precise, and that nothing is lost during the process, if a woman so wishes to believe, and if a woman wishes to think that sperm do die, surely she is lucky enough in this world to be able to think that. Then again, the sperm cell becomes something more animal. Up until today we could not have established that such a cell appears more animal, because then maybe no woman would want to have sex with her husband. And so, from the beginning of time, we as a society made sure that the naked body of a man would be looked at and accepted, because man so easily did accept the body of a woman first and woman did accept the body of a woman first because it is her body. The real-life concern of every man is not that if woman will accept him, because the body of a man is beautiful, but the concern is if woman will accept the body of a woman more. And if a woman needs to be subdued in order to take part in sexual intercourse with a man that she loves, that is not that unrealistic because she may not want this fecund statement in her body. Sperm circle a bubble like a chatter-box does. The bubble is like a space where women gossip: we have a phalo-centric society and maybe we don't want to explain the reasons why. I have no idea if there is a photographic image on the cellular membrane of sperm, *because this type of image is always on the skin of one organism that has no twin*. One day in a week will be the day spermatogenesis happens, and if that day happens to be the day of Mars, Monday, then surely the final result of this process on that day will create a army of men. One day in a week will be the day spermatogenesis happens and maybe it does not happen on a normal day and if it does not happen on a normal day that day will be governed by Pluto. *Even Pluto will extend a olive branch to a wealthy man.* Without a man there would be no baby. Men play a vital role in the processes of life itself, again, whether or not a man can decide for a woman what sex will be born from her body, whether a man can decide this for the future of his child and family may not be so easy to explain, and there is a war between man and woman, and it is a challenge for a man to decide the fate of his child, just as much as it is for a woman: if oogenesis can be accompanied by the journey of a man as well, that is up to the vital rush of planning a pregnancy, for both sexes.

   The men are at a loss, they are always at a loss because a man can cause the birth of a ovum one day sooner, but ends up achieving this another day, later. But if he does achieve sexual fertilization another day later, that will not be a normal day and if he does achieve this at the final moment, at a last chance, that will not be a normal hour or minute or second. To try and try again is a event in the sex life of a man which always must result with the ejaculation of genetic material, but to try and try again for a woman will never result in the ejaculation of genetic material. How the generic calendar is ally and enemy to a woman, the seven-day week is ally and enemy to a man.

    Spermatogenesis is not like oogenesis, where oogenesis, that tree, is limiting, where oogenesis describes the losses the life of one celestial organism will endure: before a birth the life of one existent, -who is born knowing celestial information that no one can sus out of him or her-, his or her life competed against odds, -before he or she was born-, and was supported by the constellations to always remember that some superstitions are fictitious, including loss. Oogenesis: the process of the birth of rationality. It will be a man who introduces a world to this ovum, and to this child, where a great lesson needs to be taught by a father to the child, whether the child is a girl or a boy, which is that to not be afraid to encounter loss, and to not be afraid to encounter superstition, and to overcome it, because the body of a man has overcome obstacles soaked with irrationality too. But it is the body of a woman who will laugh at the zodiac, something that seems to be introduced to this to-be egg by the spermatozoon, because the competition between a ancient procession of constellations, and the western zodiac, describes the slender incongruence in ideas between the cult of womanhood and the world's belief that the earth takes part in a wobble as it turns on it's axis. A rational woman would believe that it is not the earth that wobbles while it revolves, something that would ensure that the zodiac and the superstitious beliefs that come with fulfilling a attention to a story of the constellations where no other constellations are involved except twelve, is too superstitious, especially because we know that the moon is center to no astronomical events but travels around the earth below and above the earths equator, and does not follow the same path until the next month arrives. The ovum results from one tree of life, if fertilized by a spermatozoon, and this tree endures separations in order for the product of this tree to become a quivering cell, and there is no risk, no worry of loss in that biological process: both those losses of both polar bodies, all of those losses will certainly result in the possibility of life, if the secondary oocyte is fertilized. Surely it is in the cult of man and boys where a young man learns to not worry about loss, where a young man does not have to worry about being thrifty, because it is simply not cool, and confidence in the world of man means getting used to loss and not losing, as if the body of a man takes part in a ever present renewal, where a man is always running away and running to becoming a king, where as a thrifty woman never wants to lose anything, perhaps because if she breaks a bone it may not heal as fast as if it happened to a man. And besides why would a man worry if he needs to try again to penetrate that secondary oocyte, because it is something he can enjoy anyways! A man is just hoping that a woman will attach herself to a servitude to something superstitious, because his life is filled with more superstitions than her, and the belief in a zodiac is biblical. Because if a woman never attaches herself to something superstitious there will never be anything predictable about her, and how does a man then know what she will do next, how will a man know about the things she likes to do because a woman is surely a less habitual person than a man, but it is always the man who needs to win her hand in marriage. He hopes he can pin down as many of her habits as he can, not just so that he can know her, but so he can try to change those habits if he doesn't like them: this man is prepared for the modern day because in the modern day women today are less likely to want to stay home very often, and those women are young women too, and is the modern woman less superstitious? Nothing will change about oogenesis ever. Even if we, as women and men, decided to create informal hypothesis about which constellations the moon truly travels through, regularly, as the moon circuits the earth. And a woman will many times not be able to avoid reading her horoscope, because anything taboo, and that is veiled by the darkness of taboo, has not become popular yet, perhaps because the whole world still needs to read their horoscope. Spermatogenesis results in a living organism, one of them and then many of them, where only one of many will succeed. It is not a tree of decreasing chance, there is nothing lost from one organism alone, but it is said that many of sperm do die, and no one worries about a spermatozoon dying, but maybe it is just that they don't succeed. If some sperm do die, all that don't succeed, should we treat them like they are nothing more than dead skin cells? The purpose of sperm in sexual reproduction, in the body of a woman, in relation to the ovum, is motility, movement, genesis. The successful sperm captures the secondary oocyte and pins the egg cell down, voicing the seven-day week as a demand to obey, voices to recognize the day, the month, but never the year. To the ovum, the western zodiac is like the chinese calendar, a superstitious belief that arrived in our world on our land from the eastern part of the world, because the moon begins its circuit of the earth from its seat, lodged in the deep east, at 3 o'clock on the face of a clock. If the to-be egg was a earth instead of a moon we would never find the need to make a moon calendar. Surely if the earth is center to the universe, the moon is too. If the to-be egg was a earth instead of a moon then the greatest gift to the ovum which is a month, would not be a gift from a father. The weeks into pregnancy and afterwards are counted until full term: no longer will there be the inequality of the sexes, no longer will a quivering egg resist. To be so sexist, a sperm will conquer. Sperm along with the daily, weekly, calendar have to compete with the turning rotation of the secondary oocyte, and there they are competing with the celestial sphere, the secondary oocyte moves like the stars do in the sky, turns clockwise: every child is born with the laws of motion.

Chapter 2

Two Degenerate Eggs

It is often said, as if it is a old wives tale, that a woman sheds immature eggs during her menstrual cycle, the eggs that did not become a ovum. If a woman is real, and alive, and a animal too, this is not a stupid idea, but if a woman is also a idol, then the idea that every month she does get rid of something that was not fertilized and that did not become a embryo, is too realistic. In the risk of being petty or too detail oriented I would ask, as for what happens in the body of a female, does she really discard something that can become alive in the future? To connect two cycles of life that a woman will go through, which is the period of the menstrual cycle and the dormant part of it, which is pregnancy, we find that a woman discards nothing and carries nothing genetic. Within the usual correspondence between the period and a somewhat imaginary ovulation we have a live debate as to if some external force causes the secondary oocyte to be pushed out of the ovary to await fertilization, but we have no live debate as to what causes the period to arrive in the body of a young girl: a play between the hormones in the body of a young girl is sometimes stated, in the world, as a cause to the arrival of the period in the body of a young girl. When we are talking about ovulation in the body of a woman, we so easily exclaim to the world that this woman is subject to a fantasy that the world would like to impose on her, which is arousal, but that woman was once a young girl, and that woman in the western world will have a eternal child, and we should be afraid to say anything like that about the body of a young girl. And we rarely believe that a man is subject to such external force like arousal meaning there is no god that a man is exalted to, if all of this is true. We may know that a woman discards nothing and carries no gene. A secondary oocyte is not pushed out of the ovary in response to a external force, and besides that, the cell can stick to the outside of the ovary and await fertilization: there is nothing irrational about the body of a woman. We will find that during the life-span of a woman, a woman has the choice to believe that she is not the carrier of any gene, and when she is pregnant, she is not the lawful contributor to a embarrassing attachment to any gene in the race of mankind: if a woman is to give birth it is a humble act of giving. When we, -instead of assuming that ovulation follows the period, every time the period arrives, a sort of irrational assumption which removes the secondary oocyte, to outside of the tree of life it is a part of which is oogenesis, and makes everything that happened before the secondary oocyte entered the environment of the oviduct, inaccessible, inornate, and imaginary-, make the utter, sheer, possessive and emotional equation of the period to the following pregnancy, we recognize the vital, fecund nature of the body of a woman, and the capability she contains in her body, and then she is less of a animal that we have made her to be, whether or not we have made her to be a attractive species: it is time to be more superstitious about the body of a woman and her capability of giving birth, that a woman can plan a pregnancy and maybe she wants to avoid a ovulation that is not met by fertilization.

  During oogenesis, in this phylogenetic tree, no genetic information is passed downwards. Unlike a traditional phylogenetic tree, where lines of pedigree are long and the history recorded is ancient, oogenesis is a small inheritable tree, and is so modern. Every oogenetic tree for every woman is different. Some are shorter, some longer: some <u>forks</u> will arrive sooner, higher up on that tree, so in the event that that happened, we can conclude that the first sexual initiation transpired sooner in the life of that moon, and so the secondary oocyte is derived sooner or later as a response to that auspicious event. Every oogenetic tree for every woman is different. Some are shorter, some are longer: meaning the primary oocyte arrived earlier or later, to be circled by the celestial sphere inside the ovary. And as for the space between when the secondary oocyte is derived, and is sexually fertilized by the sperm to result in a ovum, which is not the primary fork in our oogenetic tree, it still can make a lasting effect on our tree, making this tree much longer or much shorter. It is not in high esteem to calibrate a phylogenetic tree, but for this bodily process measuring is not so wrong. Oogenesis is a crappy phylogenetic tree. To make it perfect could be a hope of a woman. After all, don't we want to plan our pregnancies as women?

There is always a feminine side to the to-be egg. As a primary oocyte, the cell is uneven. The shell around the side that we call the feminine side will always tear apart first and it is the slightly smaller side of the cell. It may easily detach or not, but whether it does or doesn't, it does not mean that the woman is more or less virile or more or less fertile, it is better to blame or applaud a fantasy for the easiness or difficulty with which the smaller side of the shell unravels. One sperm will enter through the feminine side of the secondary oocyte, *the feminine side of the cell has been exposed to air and water for so long that it almost gets soggy*, it is easy to penetrate. The spermatozoon causes the shell on the masculine side of the cell to unravel because it deposits a elastic pressure, with which it penetrates the cell, and then finally both polar bodies have been discarded, our two degenerate eggs. If our masculine constellations showed sexual prowess, they caused the feminine side of the to-be egg to separate first and then the child is a boy: the feminine side of the cell is marked with many constellations that are blue, constellations that fill that side of the sphere in abundance like the night sky in Canada. If we were to see the cell then, we would see a night sky that has little room in it for a shooting star that was unplanned for to see. If the virile constellations made their mark first, then they caused the feminine side of the to-be egg to separate first, and the child is a girl, and if we were to see that cell then, we would see a bold night sky with fewer dominant constellations in it: those stars would be shining in white but would appear farther away to the naked eye. The color of the night sky then would be a bold blackened grey and perhaps that night sky too would not observe a shooting star. But, the feminine side of the primary oocyte always left first, and so the sperm will tunnel through that side, where the skin of the to-be egg was exposed to air first. If the child is to be a boy the constellation Taurus shines brighter than Uro, or the constellation Orion shines brighter than the Hunter, or is it that during that moment, when the first polar body separated from the primary oocyte, that we were all looking up at the sky, women and men, in some philosophical quest, and it was proven that for that small moment we decided to recognize a constellation for his or her character: a prophetic glance into the future, and if we decided that we could not look at that constellation again, in another light, for any other character trait, then that will change oogenesis forever. There is a celestial competition between man and woman that we as men and women do not always understand, and there is a competition between male and female that we as men and women do not always understand. Every act we make, as individuals, as a woman, or as a man, impacts the way a lover that we have feels about us. Every man wants to conquer and so does every woman, and sometimes we go years giving in for the love of the other end of the mitsein we belong to, but when we as individuals want to extend to the world and to express our own desires, sometimes this causes us to win over our lover, *and then the virile female constellations shine brighter, if a woman won.*

      If we have no more real constellations than we already have is a mystery, though it may be true. If there is more to find out about our stars, there simply just may be more to find out about the myths of our current constellations. Based on chance and based on the belief that a birthday lies on a ancient western zodiac there are two moments of birth for this ovum and a third, the day the child will be received in the world is the third.

      For one breath the ancestors remain silent: hollowness is passed downwards in oogenesis, the sex of the child is pre-destined and no one knows yet what the sex of the child is. Somewhere during oogenesis what was hollow becomes full and only sexual intercourse alone has this duty to immortalize flesh at the site of penetration.

It is a myth that the primary oocyte can be ushered into losing the first polar body, due to masturbation, a myth that arrived to our world during the modern art era! But the first polar body is lost by chance anyways, so is it truly a myth that oogenesis can transpire this way. There is a sterile identity to a woman who has not insured her pregnancy this way and if she has not, well then, in the sky, Cygnus outshines Leda and that is beautiful too. And if the woman who has not insured pregnancy this way does have sex, will she give birth to a egg like Leda and will she, because of that myth, remain mature and not immature? Leda and the Swan is immortalized. It is known that Zeus seduced two different women who were not his one wife, and perhaps many more. But in the example of these two women, one who is Europa, and the other who is Leda, we do see that in the first case a boy was born and in the second case twins were born and they did hatch out of eggshells. These are two examples when a specific type of birth was a result of sexual intercourse between a man and a woman. Could a boy be born, or twins be born from the body of a woman if she did not plan her pregnancy and got pregnant anyways? In the rape of Europa, Zeus, disguised as a white bull seduced Europa and took her to another island and there she gave birth to a boy. And it is known, by every person who read that myth or learned that myth, that Europa was raped, the sexual act between Zeus and Europa was not consensual and she did give birth to a boy, on a island she knew nothing about, on a island that was not her home. This is a moment in history when a woman could not decide the fate of her child and could not make a wish upon a star about which sex would be born from her womb. When Zeus seduced Leda, she gave birth to eggs and twins hatched from her eggs. Instead of disguising himself as a white bull, a disguise he used to seduce Europa, here he disguised himself as a white swan to attract Leda. These two examples proclaim to our modern world that for a woman who does not insure the birth of a child, for a woman who is 'immature' in that way, she may give birth to twins and she may give birth to a boy. Though the birth of a boy is destined, the birth of twins is a awesome feat to accomplish, and is always a exception to the rule of oogenesis. And so if a woman is raped and a boy is born, that is also a exception to the rule of oogenesis. Remember, without loss, the probable result of oogenesis is a twin and the probable result of oogenesis is a eggshell.

In spermatogenesis where one lucky sperm manages to successfully coincide with a to-be egg, we feel no need to interfere.

Farming is a inheritable trait. If the early tillers of soil maintained that the primordial germ stay alive, we should want our children to be farmers and to farm on plots of land that have been farmed before, and on fertile soil that has not been farmed on before. If your child is going to be a farmer then your child deserves a imaginary line of ancestry, your child deserves to be born -if the mother wants to give the phylogenetic tree away. This child will already have a long line of ancestors at birth, but while in the belly, the constellations are the child's only ancestors and we may never know who they were but that child might know.

Not everything in a woman's body is a never-ending cycle, one cycle after the other. A woman's body is subject to so much growing and very little loss, but when does a woman exist in no cycle? When does a woman carry nothing, and still yet lose nothing? Because during oogenesis she does lose, because in the menstrual cycle, though a woman may not lose something that is alive, during the period of the menstrual cycle, woman just loses.

During oogenesis something is lost for the purpose of creating a phylogenetic tree, to create a chart where gender is compared and competes to be born, a phenomenon that we have not yet found to exist in any other phylogeny either, until now. One cell suffers losses during oogenesis, and if the characteristics of that tree, and in actuality if the very same tree, isn't inherited, It will never come back again, but if it is inherited it may come back again: a woman may give birth to a child, and then another child who was born from the exact same oogenetic pattern, *the same fantasies caused the same fork placements again*, like a garden that returns every year, like perennials. There are variations in inherited oogenetic patterns. One of them is when a primary oocyte exists inside the ovary and a secondary oocyte exists in the oviduct at the same time. In this auspicious circumstance, the secondary oocyte is older, and the primary oocyte is younger. I have coined the term Geminoid structure for this auspicious moment. This is when a woman will have a child, and then another child succinctly afterwards. This is the organic sign of a sibling, and it exists inside the body of a woman. There is a particular angle that is created in between the primary oocyte, which is in the ovary, and the secondary oocyte which is in the oviduct at that time. The angle that is created between the two cells is close to a forty-five degree angle and it reminds me of a angelic angle, it is a auspicious event inherited by the womb of a woman. The angle is a bone that lies across the skin of the ovary, much like oogenesis is a bone that lies across the skin of the ovary. This structure may be what defeated the twin: the probable result of oogenesis. If the body of a woman did not take part in rational losses, like the period, and like the dismissal of a degenerate egg, then I think it would be more common for a woman to give birth to twins every time she got pregnant. To pinpoint exactly why the result of oogenesis is not always a twin is because the to-be egg cell has a feminine side, where the shell breaks off first, which is lighter and so causes the cell to wobble, which is lighter and so causes the cell to spin, and where the spermatozoon penetrates. If a woman and the entire female gender was not choosy when the event arose to pick a mate for life, and when the event arose to become a mother with that mate, it is likely that oogenesis inside the body of a female and inside the body of a woman, and inside the ovary, would salvage every un-useful cell to make a baby out of it, and so would make sure that every growing Graffian follicle would be the cell that houses two primary oocytes, so that the cell would remain uneven throughout the process of oogenesis: the attempt to make something rational that is irrational is always to give birth to more than less to ensure that what is born good will be born for sure. To give birth to a egg is the most perfect live process.

Oogenesis is about two things, making a impenetrable egg penetrable, for when we are done with this process and we have the egg, it has no phospholipid bi layer and when we began this process all it had was a shell, frigid like a spinster, and the second thing is to avoid the birth of twins, and so if you do give birth to twins, you are lucky.

In a woman's body a moon loses it's skin and is driven into, a event no different than when Melies' rocket hit the moon.

Sperm no matter what will only penetrate the secondary oocyte from the half side of the cell, the feminine side, the face, of the moon, that we can call Diana. This can be the moment of activity for this cell that we can call feminine.

The two polar bodies are each halves of our circling constellations, completely half-moons but round. These two degenerate eggs, two shelled organisms are discarded and they leave with what is inside them, what is inside of them is information we will never understand. We could fancy a idea such as that the two polar bodies do not disintegrate, and so, what make up together, one entire celestial sphere, like the one that circles our earth, does not leave the ovary of a woman ever. But the moon cannot be center to a astronomical event forever, and two halves of a moon, once broken and pulled apart, in a celestial world, would find no reason to come back together again, and two halves of a moon, in a world full of fluid, in a world full of air, would find it impossible to come back together again, until it gets dark again in the womb, in the ovary, in the oviduct and in the uterus. After all, why was a movie about shooting a rocket in to the face of the moon so popular, as if quieting the moon is intriguing. The celestial sphere arrived to circle the moon, was consumed, and then before it could disappear it became too bright in that ovary to wave goodbye to those stars. And no one ever really asks what happened to those stars, everyone seems to want to know more about when they arrived, years after the child is born, but there is also something fanciful about discarding a organism we used, and it is even more fanciful of a idea to discard a set of information that we will never understand, and the knowledge that came from people we may not ever meet. Surely the celestial sphere does not stay, after the event, in the ovary of a woman, because then why would we need a memory of the event inside the to-be egg, and if the celestial sphere does stay inside of the ovary forever, then maybe we can never give birth to a relic like the moon.

The secondary oocyte looks like a half moon to the practiced eye.

No one will interfere with the to-be egg when gender is being assigned, because that is the time we can begin to read the child's marriage vows with the universe, so again I insist that gender is assigned before penetration occurs.

The sperm only penetrate on the feminine side of the to-be egg. The to-be egg is challenged and loses its first polar body, and then harnessed and losses the second polar body but because it takes challenging the to-be egg again, at its feminine side, in order for that cell to react, it reminds us of masturbation,- because the feminine side of the to-be egg always encountered change first-, a bizarre sort of immaturity is taken away at that moment when the successful sperm cell penetrates the secondary oocyte at the feminine side of the cell, -when the secondary oocyte is fertilized. A bizarre sort of immaturity is cancelled out, so we can forget what happened in the past, remember we still don't know the gender of the baby.

Masturbation is what causes the primary oocyte to shed the first half of its shell. This type of rule will ensure we have only a predictable amount of constellations in our celestial sphere, those we already know to exist, if we don't want our night sky to change. If we want our night sky to remain chaste. Some of us want to ensure that there remains a chastity to our night sky. There can only be so many constellations. As for the two polar bodies who's separation from the egg cell ensures the result of a ovum, they are degenerate because they did not depart in one piece. The two degenerate eggs may hold the secret to constellations we have never seen: a bizarre form of immaturity. If our celestial sphere only houses so many constellations, I think there is a possibility that when the two degenerate eggs depart from the to-be egg, both round shells are two smaller celestial spheres and there is knowledge inside of them about every broken myth, stories that are rarely popular or characters that no one remembers. And if these two degenerate eggs do exist, which means they are celestial spheres on their own, each, then it may be possible that the stars in outer space and the constellations that exist there, that we have never seen, could be seen right here in this world when looking at these two shelled organisms. They are handleable globes. They are something that we would want to look at, globes that we would want to look at and hold.

When a primary oocyte, that which is born out of one of the four hundred thousand follicles that a baby girl is born with, is formed, that moon is surrounded by the celestial sphere. If there are these almost unlimited amount of follicles in the womb of a baby girl, it must look no different than a hollow hall full of a bunch of pickpockets, that is what we see in the womb: the drawing of Paradiso from the cover of Dante's divine comedy. No woman is wrong for calling her womb Pluto.

If because of two degenerate eggs we have star clusters, it may be true that you will give birth to a child that carries with him like a backbone, the lone star, because when both polar bodies separate there is not a great chance that all full star constellations are captured as markings on the skin of the cell. At the equator of the to-be egg cell we will find star clusters. If you are lucky enough to have remembered Venus, and so she has made a mark on your to-be egg, depending on the season she may be in the attire of the eastern star, or she may be dressed up as the morning star. And if you remember her as the planet Venus, you may give birth to a boy.

There are a variety of patterns made up by the constellations that can encode what just may become a child. What is left in that egg after both parts of the crust are separated is only a memory of genetic material, it's a memory of the constellations. It is really the polar bodies that carry the genetic information and the reason for this degenerative process, which is both halves falling away from the to-be egg, is so that real genetic information can be discarded in play. To investigate more and more the reason for Dionysian celebrations: perhaps it is not so far off that we celebrate rupturing the crust of the moon, for what are all the masks for that we wear during Dionysian celebration? Are they to hide our identities while we bastardize the moon? Something we are afraid to do. Are the masks we wear during Dionysian celebration worn to hide our identities while we bastardize the moon? A celestial sphere that is center to no map! The polar bodies are discarded in play because no woman wants to lose something, it's a form of immaturity: all we really do have are a numbered amount of constellations which is a very immature thought we have.

If a woman's womb is made out of a bunch of pick-pockets: Paradiso, if all of what could be underneath the earth's crust is inside one woman's womb and we know hollowness is passed downwards in oogenesis, then the womb is a place where we would bury treasure, and if the scala naturea put a soul in the womb of a woman rather than in a mother earth...I am reminded of a painting by William Blake called The Ancient of Days.

The moon in our skies is turning a turn that the earth never sees, making the birth of a child, the birth of someone, the birth of every one of us a result of a promise that we have not yet seen but we ourselves are made of, which is the dark side of the moon.

We think of the period of the menstrual cycle as having a beginning, a first time, but we do not think of oogenesis as having a beginning but the process has a first time to resound in the womb of a woman. There is a first time for everything in the body of a woman. Only one primary oocyte can grow before the period of the menstrual cycle begins for the first time and this very primary oocyte is the young girl's dream. It is simply chasteness. Because the ovum is born one at a time, because the period of the menstrual cycle arrives first in order for a child to be born, it arrives as a signal for processes like lactation and growth, hip separation, uterine widening and elongation. For the first primary oocyte who arrives in the ovary of a young girl, it is very curious and unknown as to how long that moon has been there. When a girl becomes a woman she should know this about her body but what a young girl chooses to believe is up to her.

It is only when the young girl grows older and becomes sexually active that she begins to imagine the capability of sexual fertilization, of the type of sexual fertilization that we see in a biology textbook. The young girl surely knows that the propagation of the species lies in her womb but her concerns are always much more political: how to be nice and who to be nice to. This capability is called fecundity, and it Is illicit, and it is veiled and it Is a taboo. The day the young girl can call oogenesis fecundity perhaps may be the day that a young girl becomes a woman: but This is a taboo. Then there are false pregnancies, fears and there is no more of a young girl, because a young girl's greatest dream is to never have a fear. When a woman was a young girl she was living her dream. She was living a life that she could live outside of all obligation, a singular lifespan. But to hop, skip and jump from being a young girl to a mother is just not fair to the young girl, who though never wants to become a woman, and it is her greatest plight to prevent that from happening, when she becomes a woman at least she can hold on to a thread of herself left from childhood, and this is that though she may no longer be a young girl she can still be a woman and remain a virgin. Virginity is decadence to a young girl, and when she is subdued by no one but herself, her dreams are comparable to no one else's. Something she almost relishes. And when she knows that fated idea of a priori, that who knows if the menstrual cycle came first, before the moon arrived in her womb, before the celestial sphere came to circle the moon, *before a celestial sphere stood there alone*, she has already fought the religion of science.

It is harder for a woman to fantasize about a woman. This does not make the woman any less valuable for fantasy but incredibly hard for a woman to fantasize about in the first place, giving what is woman a numerical value, a identity based on cost: how much does a young girl have to give, to measure, to agree, to accept, that the body of a woman is equally attractive to her in a phalo-centric society. Perhaps it was a young girl whose mind fertilized the modern-day pin-up woman, the female pop-star.

Could it be that acknowledging herself the young girl spurs on oogenesis, only to make her desirable for suitors, suitors she will never respond to given the circumstances of her virility and her frigidity, but this making her the most desirable woman in the world.

On the hierarchy that is headed by man and woman, descending from mythology alone, where there is no one god as head, where man and woman are two gods, the menstrual cycle of woman, of the first woman named after herself, the one woman named woman, is a cycle of life which is no different than any woman's who came before her, but is however unique in a peculiar way, because all of the other women prior to her birth were mythological women first and she was a woman first: her priorities lay in what made her a woman. Her menstrual cycle, the entire wheel of her life, for this is what it is for her, it was, and it is and will always remain peculiar. We talk about the menstrual cycle of a woman, we manufacture products specifically for the use of a woman during her period. We never did that for mythological women: this woman that everyone is talking about, who all of art history tried to describe, is a woman who lives in the industrial world.

The fork in oogenesis, the first fork, The fork of oogenesis, -when the primary oocyte becomes a secondary oocyte,- allows for subtle differences between the pregnancies of women. A primary oocyte can be formed before or after the arrival of the necessary period, the period necessary for a first pregnancy. How many periods should a young girl have before her womb is taught enough to carry a child? Is it one, or many? Surely there is only one necessary period for a pregnancy. If one ovary begins to work first, so only one ovary takes part in a menstrual cycle each month, meaning that even if no ovary takes part in a imaginary arousal and even if nothing that came from the origin point which is inside the ovary is discarded during the menstrual cycle, and even if only one ovary will give birth to a primary oocyte that will become a baby, a first pregnancy need only one period. I believe that the period brings darkness to the ovary involved and to the both oviducts and fallopian tubes and to the uterus. During that time there is light in the other ovary. And so, when the menstrual cycle began in the body of a young girl, there was darkness in the ovary involved and light in the other. There is always a necessary period for a pregnancy, only one. And if there have been many periods in the body of that woman before that, the necessary period is different. If ovulation did happen at any time in the life of a young girl or a woman, it happened after the necessary period. As for the story of the two degenerate eggs, as derived from oogenesis, the two polar bodies are discarded in play of course, but if a primary oocyte is formed in the ovary of a young girl before the menstrual cycle arrived for the first time to the body of that young girl, the formation of the first polar body is a different sort of play, like some sort of recognition for that young girl. For this event to be something monumental, something that even history could not silence, and something even history would speak about, and remember, and echo throughout the ages, -behind the scenes-, something that came straight out of a comedy: perhaps that young girl thinks she devoured half of a moon that is in her ovary, herself, through actualizing her own fantasies, because she did not lose it to her first period, because her period has not arrived to become a permanent part of her body yet, but then there is surely a secondary oocyte which has been pushed out of the ovary and will be subject to the world of the oviduct, but was there no conquest there? If we were to fashion a identity of the lover of this type of woman, everything could become tragic for her: *this could be a tragedy for a very happy and very funny person.* Maybe that type of girl has a lover who is spiteful and vengeful and jealous, *and then he wants to be her world.* Because no egg is discarded during the bleeding of the menstrual cycle, it is almost as if that young girl was successful at the conquest over herself, her mind and her body, because she bled out no egg that was there first, if the primary oocyte, that moon, arrived in her body before her first period, because the very first time that young girl tried to conquest over herself, she changed the world in her womb. If the primary oocyte does not arrive before the first period in the life of a young girl, then there is never really a worry about anything having been lost when the period arrives for that young girl, but then there is no conquest on behalf of a young girl. But, there is not always a primary oocyte in the body of a young girl. There is not always that cell there, housed and adored. There is nothing wrong with soma. Even though the primary oocyte and it's precursor is always a sex cell, and is always feminine, somatic cells within the ovary do exist, we just don't have to believe that a sex cell is derived from them.

The major fork in oogenesis happens at a defining moment when a primary oocyte becomes a secondary oocyte by losing half its shell. The moment before this happens, as the moon is waiting in the ovary, may last a long time, and from when this event transpires and afterwards, - from when the primary oocyte becomes a secondary oocyte and afterwards,- the cycle of this moon is allowed a subtle break, a silence, a pause, a wait, because that secondary oocyte may wait for sexual fertilization for a very long time, and the length of time of this break may last as long as it took for the primary oocyte to be formed. This fork ensures that a primary oocyte can be formed before the arrival of the first period, *because the primary oocyte will remain safe in the ovary, no matter what is happening in the rest of the womb and oviduct, no matter what muscle tension, no matter what muscle spoil, as if it is harder for a to-be egg to take part in a disappearing act inside of the ovary, because there is nowhere for the to-be egg to go.* Oogenesis is literally a bone lying across the skin of the ovary, and is also a pathway that the primary oocyte will take, to become a secondary oocyte, sticking to the outside skin of the ovary, so surely, like what is characteristic of a fork in the road, a major fork in oogenesis can also prevent a event from happening: it can prevent the primary oocyte from exiting the ovary. When a secondary oocyte is available, this is when the necessary reality of the reproductive imperative is important to every family member of that young girl. And it is not to say that the biological clock of that young woman was in question by her family and lover, and if it was, it is not to say that there was a worry or concern of time running out, and if it was, it is not to say that that young girl had outlived her fertile years and had a biological clock that was ticking away, but, blood and skin are shed from all open pathways of the womb of a woman during her period, and only the primary oocyte is protected inside the ovary, always. If a moon arrives before the first period in the body of a young girl, the celestial landscape is no different for that moon inside the ovary, even though that moon will live through at least one period, but who is to say that the landscape there,- like the constellations in our night sky are a terrestrial landscape subject to change, only according to our view, depending on the weather-, is not different during a period, and because the first period did arrive after the moon arrives in the ovary, it is as if the lining of the ovary, that thick skin and it's protective barrier is also signified by the major fork in oogenesis. If the moon arrives after the first period did arrive, there is always a worry that the period of the menstrual cycle will do away with everything, because outside of the ovary no to-be egg is protected: there may be a notion that if the to-be egg, the primary oocyte, does arrive in the ovary of a woman after her first period arrived, then Everything will happen faster for that young girl, and then again will there be a worry of the family and the lover of that young girl, that her biological clock is ticking away, that there is less time to count every one of that young girl's fantasies to make them come true, *because she matured later*. Because no egg is discarded during the menstrual cycle, we could imagine that the environment in the oviduct is unique and sustains the secondary oocyte despite the arrival of the period, and if the secondary oocyte sticks to the outside ovarian lining and anchors there until a successful sperm arrives to meet the to-be egg, then perhaps the secondary oocyte can watch the oviduct lining fall away along with the lining of the rest of the uterus during the period, the view of that moon is a other world view.

Perhaps following that far into this biological complex would lead us to question instead why there is a clan for women, and why there are planned pregnancies as displayed by those taboo cultures. The answer to this scientific reverie is simply that the family and the lover of a young girl is always concerned with her reproductive capability, and no day goes unchecked where that family does not know about her sexual capabilities: we must be suave and swift to take advantage of such a auspicious moment as the result of years of waiting, which is the secondary oocyte. And if a young girl simply wants to believe that the two degenerate eggs disappear or disintegrate, she can, but could you believe that a to-be egg can disintegrate if sexual fertilization does not happen? Perhaps a to-be egg taking part in a disappearing act is not unbelievable.

The ambivalence still remains in a peculiar period, which is that a primary oocyte may exist in a woman's body for as long as she wants it to.

The key thing to remember is that the primary oocyte is not discarded during the period, and there is one beginning to the menstrual cycle. Like a clock, it was set up to run once, and go on forever: this is a brief moment when we can remember that there is one menstrual cycle in the body of a young girl and in the body of a woman, that both ovaries do work as one unit, even if they started to work at different times.

Why is it that one moment of arousal can cause the arrival of one primary oocyte in the ovary of a woman, and the small to-be egg will follow into something that will result in life eventually? Because, that is the ideal. Because a biological reality such as this, will cause the family and clan of that young girl, and the clan of womanhood itself, to plan for the pregnancy of that young girl. Because this is what created a planned pregnancy, a rhythmic birthing pattern amongst women and ancient women, to fathom success and avoid loss and failure. After the primary oocyte is gone, and the secondary oocyte is born, the life of the to-be egg exists outside of the ovary. There is surely room for another moon in the ovary, but only one. The secondary oocyte has a life of its own, sticking to the skin of the ovary, but that cell still has half of a shell, and now every day counts. The stars that did not shine the brightest did not make their mark on that cell, were not victorious, and in the oviduct the secondary oocyte is alone, and like a dizzying spell no one knows where the stars went but if you can see that lonely cell, the masculine side of it, where there is still a shell, has been engraved by the weaker stars, as if those stars were blue when they shone, and not yellow or orange. The masculine side of the secondary oocyte is like a lesser part of a tablet, engraved, but hardly readable, hardly audible: those weaker stars were of a lesser magnitude. The secondary oocyte has its own cycle, and it spins, and it will solemnly wait until the successful suitor arrives. When the secondary oocyte is pushed out of the ovary, with momentum, it spins, because the cell is uneven and so the lighter side, the feminine side of the to-be egg causes the entire cell to spin. If a chicken's egg is put onto the table it will fall to one side, the lighter side of the egg. The feminine side of the egg causes it to fall to one side and so does happen to the secondary oocyte. The spin of the secondary oocyte is a revolution, with a compulsive tendency it will be pulled, by some romantic desire, to stick to the side of the ovary eventually, if the cell is not fertilized before. Not much will happen to the to-be egg while it waits, and with it wait the weaker constellations, those blue stars and their homes in the sky, outlined on the masculine side of the to-be egg, they are waiting quietly for one spermatozoon to warm the internal core of the to-be egg so that their message that is engraved on the part of the shell of the moon, that has not broken off yet, will be heard. Upon sexual fertilization the second polar body will be formed. What is left of the shell surrounding the masculine side of the secondary oocyte will break off and the stars will leave their mark on the skin of the masculine side of the cell -which is still un-marked,- when the to be egg is crisp again: for the second time the shell will become crisp and the message from the stars is branded onto the skin of the cell. The cycle of the birth of the ovum which begins at the formation of the secondary oocyte, is the second step, the second part of the process of the menstrual cycle because one period should have arrived before the ovum is formed, the ovum will always come after a first period. If one child will come soon after the other, to arrive in this world from the womb of that woman, then the cycle of the birth of the ovum, which begins when the secondary oocyte is formed, will coincide with more fantasies and a successful woman will know which fantasy caused the birth of a moon in the ovary, again, while, the secondary oocyte is awaiting a big change: no amount of exasperation will cause the shell of the secondary oocyte to fall off on it's own, this is a time if a woman does want to get pregnant, where a suitor is vital. And a empty ovary is not far away from a blessing. And the menstrual cycle is the first step after virginity. Some women may not want to believe in ovulation because the values of such a concept adhere to a superficial reality where the connotations arrive that the body of woman is subjected to unforeseen arousal. There is still no reason to believe that that type of arousal is connected in any way with the reproductive imperative that the body of woman is aligned with in our society. So then, we can easily say that the resulting secondary oocyte can be formed before the arrival of the period, due to fantasies the young girl is involved with, anyways oogenesis is a pedigree of decreasing chance so it is not unlikely that every event in the reproductive capability of a woman will happen in accordance to some slim chance of hope and fortune: with a strong link to ancestral mothers a young girl can always be successful when enduring every one of her own fantasies, and then finding a strong suitor, someone without who that secondary oocyte will not be fertilized.

There is a period of time in the belly of the young girl where there is no hesitancy to adorn no cycle, at that period of time the young girl simply has a soul. The young girl has a soul first. Her period arrives without hesitancy soon afterwards, but it is as if that soul of hers is wearing white raiment from a el Greco painting: the primary oocyte is much too far up, towards heaven, to be tossed out with the corpus luteum every month. If the young girl becomes pregnant and ends up giving birth to a child, when does she gain another soul to give? Inside the womb of a woman a miracle happens because according to the law of nature, where something that begins to grow should die with certainty, inside of the womb of a woman, it doesn't. Life is so certain in the womb of a woman that we can estimate a due date for pregnancy.

Oogenesis is a degenerative process because the closer the to-be egg gets to becoming a ovum, it endures loses. It is a pedigree of decreasing chance because as it grows, it grows closer and closer to the least amount of chance it could have, to not becoming a ovum. But truly oogenesis is only degenerative because there is no other name for the two degenerate eggs that play a vital and still unknown role in the process, the two polar bodies.

One sperm cell is what will make our moon something animal, whereas before it was something pristine. A moving film about how a secondary oocyte becomes a ovum, a moving film of how one sperm is successful, is a modern day dilemma, because it was the spermatozoon that made a ovum a animal. Man and woman are two different species, or what is yet to be seen, but when we look at the reproductive essentials of both man and woman, we see that one cell, the one that is in the body of a woman, can be viewed from a lens that is much more progressive, futuristic, ancient and emotional, and when we look at the spermatozoon it is hard for us to see anything other than a animal. No sperm cell is racing away from death. If anything is alive from the two sex cells, surely it is the spermatozoon. The secondary oocyte is so cold from the memory of the heat of blue stars, a memory that dances on the masculine side of the cell which still has a shell. These blue stars sustained cold light, to keep that cell as cold as possible so that the spermatozoon is successful at the right time for it, and a man will wait until a woman is cold, on the outside as well as the inside, so that everything that he wants to achieve which is to make his lover a wife and a mother, will be achieved, because when that secondary oocyte, the coldest cell of the body, has progressed to becoming the coldest cell of the body, then the skin all over that woman will be cold too. There is no celestial attribute to sperm except the day in a week and in a month that it burrows into the secondary oocyte. Something has to be real out of the birth of a ovum, something that we can see with the lens of a camera. Something is alive. Every snapshot of this moment, of sexual fertilization, will prove that only one thing was alive for one moment as the spermatozoon caught a to-be egg just before it died. After everything the to-be egg went through, it is not ironic that the cell would die, -especially in a world where we fathom death much more than we fathom life- if the hands of a man would not save it. Most of the world does not want to know celestial information, most of the world would not wait to hear it, but some lucky women and some lucky men did want to know and did want to see it come to life, and with knowing this they have had children.

I believe that the primary oocyte and the secondary oocyte, and just as the cell becomes a ovum, is the coldest cell in the body of a woman. The cell remains generally the same density throughout the entire process of oogenesis, and so it stays generally the same temperature, which is cold, during the whole process. The shell around the cell can get hot very quickly, even as hot as fire, and become cold again very quickly. Because the cell remains generally the same temperature throughout the entire process of oogenesis, it is a reminder that one cell takes part in oogenesis at a time.

In a alternate play the corpus luteum is discarded. The corpus luteum is the only organ out of the two species, man and woman, that is discarded and grows again spontaneously, in the body of a woman. The corpus luteum is the sun, a circular, wide, shining sun, and this is the only time that we can watch Diana, and Dionysus die, two textured pearls are claimed by the clamshell: it is a old wives tale to believe that both polar bodies are claimed by the corpus luteum, as if a idea like that came out of mythology, there is no need to prove it wrong because it is beautiful. In the body of a woman the creation of the two polar bodies is the only time in the whole universe that both Diana and Dionysus can be two separate moons.

The corpus luteum that comes out of a woman's body was never planned for, not when the woman was a baby in the womb, not in any genetic code. We honestly thought that two penetrable ova would be the result of the activity that first began in one follicle at a time in the ovary of a woman,- but that would be true if oogenesis was not a phylogenetic tree where forks and a major fork is part of the design of the process, if the beginning and prehistory of a to-be egg was not meant to be mapped. It is a old wive's tale that the corpus luteum chose to consume one of the two penetrable ova -who could have been the result of oogenesis- to never give back. We wanted to believe that the period caused a finite end, every month. We wanted to eliminate the irrationality of the period.

Part 2

Chapter 1

The Intricacies of Nature

All of life has already been classified in a chart of life that has been kept secret by nature. I have asked nature about this chart of life during the night, while everyone was asleep, only to find out that it is not vain to try to paint one from the Canadian perspective. If we want to, through taxonomy, classify all life, we will result with a picture, with one grandiose image and in that image all of life has a history, and all of life has a present: it will be a moving picture like the scala naturae. It seems much more rational that animal life, and plant life, and all of nature, like us, arrived on this earth with phylogenetic trees that only compare the success of a animal as versus their archetypes. To compare animals and all life, and to do this in a chart of life, takes dignity and divinity out of that chart, and then we are left with a phylogenetic tree which will try to explain the evolution of species, when our concern is to not sublimate the origins of all of life to one primal species but to rather make a modern ladder of life. There have been prominent figures amongst all natural life, like Van Gogh's cypresses, like Lawren Harris's iceburgs, like Emily Carr's pine trees, like the beasts of the bible, like the horses of the bible, like the dogs and cats and other animals that have saved people's lives, like the animals of a totem pole, like the dahlia of Givenchy, or the daisy of Marc Jacobs, or the roses of Chloe. Like the rivers that have names that people have crossed or rowed down, like the owl of ancient Egypt, like the cat of ancient Egypt, like the infamous animals of who's masks we have worn. We see a history for nature that has been documented by art and culture and science is not sure what to make out of all these facts. Uro the female bull always presents a opportunity for us to fall into a story of a life of a animal that reminds us of something prehistoric, and her cave drawing can make us believe that there was a prehistory, where her life was real and not mythological, primal, old, ancient but not unreal. The cave drawing about her is a piece of art and we Do always come to claim it in the modern day, so it Is modern, but sometimes we can not help but fall in love with a idea that she is the oldest animal alive, as old as the bisons. After all of this history that the animal and the plant and all natural life does have, in our world, it is undeniable that when we include a animal, and all natural life, into the chart of life that we will make, that the myths and stories about that animal will show in the painted face of the animal. And so it is only man and woman and the face of man and woman which we cannot include in this chart easily, without some reference to history and art history and mythology. I think it is fair to say now that if a man and woman do make it into a chart of life painted in the modern day, that all animal life and all natural life assisted man and woman to do this. If a man and a woman do need art history to be accounted for in a chart of life, if only simply to reassure in a painting like this, their very existence, or be it because a man and a woman both hope that the inventions and the ideologies and the dreams that they made, will be remembered, then maybe a painting will make it into the chart of life, as a invention of man, as a idea of man, as a representation of a part of history and mythology and the modern day which perhaps made the world better. And if a painting does make it into the chart of life then a statement about mythology will be made, that our myths did impact our lives because many good paintings have been painted about myths. If a painting from art history makes it into a chart of life, nature itself and the kingdom plantea and the kingdom animalia, who will surely be in this chart, may not ever want to explain what this painting means and all of nature itself may not ever want to explain what this type of idea means and why it is in the chart of life, so we will just have to figure out why for ourselves. Because we just may have the luxury to create this chart, who's to say that a simple grouping of objects shouldn't be in this chart, like a group where a map and a clock would exist, like a group of musical instruments, like useless things that we would find in a dollar store and useful things we would find in a hardware store. Would that be too much to ask for in a world where there are many things? If these objects have a strong enough history that they have a body of identity trailing behind them, like a history, like a gender, then they should be in this chart. And as for objects that were manufactured by the hands of man and even became sellable, - like a food invention,- if they do make it into the chart of life, the kingdom animalia and the kingdom plantea will explain as to why these objects did make it into a chart of life with all natural life and with man and woman because house pets did benefit from them too. Biblical concepts like the firmament above the firmament should be in this chart, according to many Christians. Should we leave it alone because it is almost forgotten? A kiss should go in this chart. A little black cat named Rak says: *a bird should go in this chart!* Nature may remain quiet so as to not tell us the date that inventions were made, when we are deciding what should go in this chart of life: that these inventions may not have been made at the times that we think they were. As for the life-spans of animals. It is sacrilegious to think that animals have not lived alongside us since the beginning of time. Many animals have lived a very long time.

The scala naturae is a hierarchal classification system of life for all of the earth and the universe and its species. When a animal enters a chart of life it is not easy to remove his or her alignment with what it means to be a species, by this act, and the same goes for the rest of nature, and so the Scala Naturae could not extinguish what it means to be a species for natural life, even when including natural life in that chart, with god, and while putting god at the top of that great chain of being. Some would assume that in that chart a assumption is being made as to whether or not some species are more intelligent than others, and so the higher a species does sit on the ladder of life, the more prominent and the more intelligent the species is, but I do not think this assumption is true about the scala naturea. I think in the scala naturea the stories of mankind were depicted as being far above the earth's curved surface, as compared to where the other species sit, on this ladder of life, because through that religious perspective no one knew where mankind really should exist in the chart of life, especially because the stories of mankind are wrapped up in the rational or irrational circumstance which is that men go to heaven or hell according to their actions and their devotion to god, and that men should want to go to heaven. In the modern world when we may have found that heaven and hell do not exist so far away from our daily life, mankind will take a place in the chart of life next to the species of nature, somewhere still above the earth's curved platform, - because there will be a hell in that chart of life and it exists underneath the earth's curved platform. Even though animals do not go to hell, the underworld and the role animals have played in the Egyptian underworld and the Greek underworld can make us wonder what animals do know about hell.

In the scala naturea no one was alone because god was at its head. The chart told a story. It was a moving picture that is our lives. It was a chart of life that came out of a religious perspective but however buoyant it was, however animated, however dramatic it was, it was a better perspective than what the scientific world could conjure alone, without religion. Whether we believe in angels or not or whether we believe in god, the scala naturea told a story about life because the drawing suggests motion: the image, the scala naturea, the great chain of being, suggests motion because man is shown, in that picture, as traveling between heaven and hell. Also, motion is suggested in a chart of life, once we incorporate the earth's curved platform in that picture. And motion is suggested in a chart of life once we incorporate the celestial bodies into that chart. Any chart of life that we can benefit from will tell a story of life, *it will look like a moving picture*, because we are living and are living within earthly, planetary and celestial seasons. In the scala naturea we see celestial bodies that are alive, and animated. Surely today we can benefit from a image like that when we attempt to make our own chart of life from the Canadian perspective. If angels are real, we do not know what species they are, much like woman, but the angels made it into the great chain of being, and I have heard the angelic choirs sing so if the angels are in our chart of life it may not be so wrong. In Dante's divine comedy there is a chart of life which describes the different angels, it is called Plan of Paradise, and so now we know that there are two distinct charts of life where angels do exist, so perhaps that too will give us another reason to include them into our new chart of life.

      In the present day, in the modern day, where animal life has been involved with all of our inventions, when we can finally imagine a industrialized world, man and woman remain to be married by a divine law. A man who came out of history and mythology, and a woman who came out of history and mythology must be in our classification system and these two must become pinned down by gender. Through the relationship between man and woman, -which is contingent, secular, common-law, married, sanctified,- that which we are still waiting to accept into a chart of life, and only once we see a declaration to the universe that there is a resonance in that relationship of a loyalty to gender, gender being the first biological classification system, then, once accepted by nature itself, man and woman will introduce to this chart of life, a law that natural life always only accepts as a horray! Then, a successful endeavor will be achieved by man, and woman, which is that the kingdom of mankind made it into a chart of life, and found a place to exist, inside a natural order and not outside of it. Throughout all natural life gender is a natural occurrence that is signified clearly and concisely and no animal will have us compete with the rational reality of gender, to win, and fall, and lose: within man and woman, we are wrestled down by the requirement of nature, which is to recognize our duties as man and woman, and nature will remark if we confuse our gender roles, and so the play between man and woman can be dramatized on a theatre like a biological classification system of all life, where nature is our humble audience. Nature is our humble audience, as we, man and woman will always, on that chart of life, stand rather precariously, something that can summon laughter from the natural world, but yet we are humble because it was our destiny, in the face of nature, to finally, extinguish the confusion in gender roles.

It is a classification system that submits gender to its will. And a woman is often running away from femininity and the feminine cell.

Man and woman are the Names of two existents. Mankind is the species to which both man and woman belong, but for a woman, there surely is a sphere of culture where she belongs alone with other women and if we can call that sphere a cult of women, then so be it, but it is yet to be seen if a group of women, together, will make it into a chart of life, as if the Amazons could fulfill a part of a chart of life where in that space there is no need for a man. Female angels made it onto the ladder of life, they fulfilled their own rung on that ladder, and they were called thrones. Maybe if we were to include the Amazons onto a chart of life, then their gender opposite on that chart would be to include the Olympians, a group of strong men, as well. But if the Olympians make it onto a chart of life it will surely be because the cult of woman was strong enough to make it into that chart first. If a man and a woman do make it into a chart of life, as mates, which surely they will, only the addition of angels, to this chart of life, will provide a escape for that one woman, into another species of her own identity. The animal world will only really embrace man and woman, into this chart, together, as one species. It is undeniable that a woman belongs to her own group and culture, with other women only, but in the words of Tigerclaws, a female cat: *a woman will be humble to join the race of mankind in a chart of life, or else nature will not be humble to incorporate her alone into a chart of life.* Nature itself and the many kingdoms of species will usher in a time where a man and a woman can become viable entrées for the chart of life: and so both man and woman need to be accepted by nature during some period of time in history in order to make it into a chart of life. I feel that using the Christian tradition as a sounding board for all ideas, like for example marriage, both man and woman will be successful when facing nature. But a man will always need recognition from a woman, of his species, as well, to enter a chart of life, because every species of the female sex will judge him then and there, unlike how a woman is judged by every species: in the animal world motherhood is a tribe that every animal belongs to. Because of two phylogenetic trees, built on language, and built on a lineage of ancestors, both man and woman have a intensely intriguing biological origin, which is their names. Man and woman were not born first but the inclination to always want to put man and woman on the top of a chart of life is undeniable, but I have come to the realization now that this action is our cry for help, to the world of nature, because there are so many species within the kingdoms of life, and we want to put them all in this chart, and we want order in this chart, because that is what nature demands in the first place. *But man and woman were not born first.* If we were to give a date for when man and woman were born it would be somewhere in the second half of the twentieth century: Chagall's <u>The Birthday</u> painted in 1915, when Dionysus returned to give life back to a woman, with a kiss. Art was the only realm out of the arts that documented the development of a mankind we now know to be young, and this was done through painting primarily, this was done through the painted images of man and woman. But it is now that we know that mankind is young because those paintings of man and woman have always been youthful because modern art died, but modern art died and kept the idealistic vision of a man and a woman alive, because before it's death the last image of man and woman, though was idealistic, was not true to the form and function of the body of a man and the body of a woman: the image of man and woman was not realistic just before modern art died. The modern day man does not need the woman he loves to come from history and mythology, thus the response of a man towards a future we have not yet seen, thus the drive of a man to make a home for himself and his wife and to make a future which involves children that came from him, thus the drive of a man to maintain a sense of normalcy in life: a man will much more easily take a vacation away from mythology to make a home for himself and his family. A man does not need a primal myth of what is a woman in order to include her in his roster of gods. That is a very modern day fresh-faced approach to worshipping a woman. That type of man can make modern art, and when he in reverence to the woman he loves, makes a sculpture of her, he does not have to make a real-to-life image of her in order for him to believe in, in himself, that the idealistic sculpture he did make of a woman was more than good enough at expressing his love for her and his worship of her. Including her as a companion in work, alone, is a sure sign that a man can worship a woman god. And so, it is a man that caused modern art to die. We could have and we would have given birth to man and woman again and for the first time in a modern era, but when I think of Brancusi's Kiss painted in 1907, I think we did. Because of all of this art making by the hands of men and women, and because the subject matter of a lot of art has been man and woman, man and woman are viable entrées for the chart of life, right now. Perhaps it was during the renaissance that we should have given birth to a classification system that incorporated man and woman, but at that time man and woman were not born, they were born again, but not for the first time.

The gender we have found to exist in, for example, a natural body of water -like a lake- cannot change and will not change if there are strong enough myths to support why we understand the object or piece of nature to be male or female. We want something to be true about the gender of objects and natural life that we can't change. Gender has become a category, specifically two categories, male and female, because these two categories are prevalent in the living world, it is almost as if the categories came first. In a modern chart of life we will divide the chart by gender. In the biological sciences we can no longer proceed without noting what gender we have found, of the jellyfish in the ocean, of the mountain goat, of a daisy, of a wildcat: we have to know the gender of every species. We are related to all species by the category of gender. For simplicity's sake, in order to classify all species, we do so by gender, not only because it would make a very simple system of classifications, the reason is more deep, it is because we are searching for why we are attracted to dualities. Though the scientific method distances itself long and far from mythology in many ways, it makes a place for mythology, and it is where the species that are found are named. A turtle dove comes from our myths. A hydra came from our myths. A apple came from our myths. But like heir and heiress, like lion and lioness, in places we have sexualized proper nouns of species. We have done this with many species:

the horse and the mare, the cow and the bull, the chicken and the rooster, the dog and the bitch. These are examples of the titles of the male and female of a species that contribute goodly to a chart of life. There are not very many examples like the lion, or Uro the female bull. Why is it that the species do not have a history that we recognize alongside ourselves when many and most animals have lived as long and as old as mankind? What this means is why are the myths and the stories about the species not good enough, not history enough for all species.

In our biological classification system, we need a heaven and we need a hell. And we should not forget the need for a old map and a new map, and the generic calendar and what is a basin and who is the owl.

There was a time in history when some very important people used a map to hunt for treasure. These people were gods. They did not forget about heaven and hell, and how important it is to be a part of a chart of life. They did believe that heaven and hell should be a part of a chart of life too. In our modern day we can easily say now that these people gave birth to the mythological age which we look back to, in the western world. If we in the western world do want to remember a past, we do not want to remember any other age, we do not want to equate any other age with our past other than the mythological age. Any other age did not benefit a woman as much. The stories that come from that time are a lot more fulfilling to a woman than the stories that came from a displaced fertile crescent. If we need a pre-history, surely the lives of the ancient Egyptian gods and the ancient Greek gods are pre-history much like Iberian masks are pre-history to the artist, like the blarney stone is pre-history, like the totem pole is pre-history, like the Venus of Willendorf is pre-history, like the cave painting of Uro the female bull is pre-history. Women in the western world are lucky that we have a real mythological era. Minerva is pre-history because she was a goddess that even the ancient Greek gods like Athena worshipped, and Athena named her ally, a owl, after this goddess. *There is a sculpture of her, clad, like a soldier, in green bronze, she is a spirit.* And when we remember the mythological era that these two religions,- the ancient Greek religion and the ancient Egyptian religion-, made with their own hands, then a woman will find that the archetype of the saint is not a attractive and beneficial archetype to be forever, if you are a woman, because these two types of people have caused a woman in the western world to inherit wealth from them and she wants to take her wealth with her to the afterlife. Their artwork is our expensive object, like gold or silver and we as women in the western world want to take their objects to our own afterlife.

In the western world we have inherited stories and traditions from Greek mythology and Egyptian mythology. This is because both these mythological religions worked together to make one out of the firmament above the firmament and the firmament below the firmament. The firmament above the firmament can be thought to be the heavens, the space that exists outside of the farthest earthly atmosphere where the heavenly bodies are housed, but it can also be what we think of to be a heaven, another flat ground above our ground and above our plain view, Mount Olympus for example. But because we can view past our farthest earthly atmosphere -during the day we see the sun and at times the moon, and during the night we can see a unending span of stars and celestial bodies and the moon-, we don't know where heaven is. Most of us refuse to believe that the heavens is heaven, and that idea is derived from our mythological religions. The firmament above the firmament can also mean earth, which means that there is a hell underneath. The firmament below the firmament was never mentioned in the bible other than having been alluded to in concepts like hell written in the new testament. But the firmament below the firmament can mean the surface of the ocean, because though we compare the earth to sky and water to land, the water can also be compared to the sky, -in a mythological story- like a piece of land, because the oceans are so vast. There is more water on earth to contrast with the sky than there is land, there is more ocean, and that is really where the concept of the firmament above the firmament began: the ocean top can be the firmament above the firmament, -perhaps we were all meant to walk on water. The firmament below the firmament can be the ocean floor, underneath the surface of the ocean, or a cave hidden somewhere underneath the ocean, or simply the land mass we call the earth, our soil who is underneath our vast skies. To the ancient Egyptian gods the underworld was nothing to be afraid of and nothing bad, and to the ancient Greek gods the underworld was a place where you could go and come back from, depending on who you are, and in their faith, the underworld was fire and brimstone, and only sometimes a cold place. To the ancient Egyptian gods the underworld was nothing bad and nothing to be afraid of because animals could go there. According to that tradition, animals did go there, like the owl. That place was not as warm as the hell we have known to exist underneath the earth's crust. We haven't found the firmament below the firmament on the surface of the ocean but we've sailed the ocean like a sea looking for that location. Both the ancient Egyptian gods and the ancient Greek gods believed in a underworld, instead of hell, but both these traditions made the concept of a hell intriguing for the western world, and if it were not for these two traditions we probably would not have been able to face the assailant the devil, in our modern day. The only underworld Poseidon can be king of is beneath the ocean, and if this infamous underworld that both these ancient traditions believed in is underneath a ocean, this cave is surely where the river Nile meets with the Mediterranean sea, and Poseidon reigns over that place too, a place that the ancient Greeks call Hades. Both these traditions believed that to get to the underworld one would have to either cross or row down a infamous river. On ancient Egyptian papyrus we can see a story that the Egyptian gods painted about a possible exodus of the Egyptian gods and the Egyptian royalty, from their land, to another land. In those paintings the river Nile was always beside depictions of the underworld. The underworld was not often depicted, in that tradition, as dark and scary but was sometimes depicted with light and natural light. From this we can deduce that, -because on a map of the world Egypt was always underneath, was the south of the western world-, at that time it was believed, and maybe even now we can believe, that Egypt was the only underworld we have, even though *the ancient Egyptian gods did want to leave to get to the underworld*. And these gods lived beside the river, to the west of the river so if ancient Egypt is the underworld on a map of the world, *as the whole world sees Egypt*, and as the ancient Greek gods saw Egypt, then the ancient Greek tradition that the river Styx is to be crossed to get to the underworld could mean to us that like a x marks the spot where treasure is buried on a map, the river Nile is a valuable marker on a map that a treasure hunter would use to find buried treasure, or a valuable place where there are valuable things. *Perhaps the entire Egyptian religion was wronged when the river Nile became the river Styx, because these people had to leave their land to get to the underworld, and every river is valuable, every river is a valuable resource for a nation.* But the Nile became the river Styx anyways. When we uncover ideas of hell and the underworld from ancient Egypt we always seem to find the idea of a hell existing beside us, a world, a underworld existing beside us rather than underneath the earth's curved platform. That hell can exist alongside our world is a ancient Egyptian theme, because on a map of the world the ancient Egyptian gods lived beside the river Styx. To a industrious world, if the underworld was anywhere under the ocean, it would be in that part of the ocean where the river Nile, -which through exchange of religious ideas between ancient Greece and ancient Egypt, during the mythological age, was agreed upon to be the river Styx,- meets the Mediterranean sea. If the underworld was anywhere it would be in that part of the ocean, beneath where a river meets a sea. Because the ancient Egyptian gods were industrious, and they did want to leave their land to avoid doom, - to avoid the failure of their farming communities-, they decided for themselves that the Nile could be a mythological river rather than what was then a modern day valuable resource: they adopted a ancient Greek tradition which was to want to take the route down that river to another life. The only thing that made these gods different from the ancient Greek gods, and what could make us believe that they were younger than the ancient Greek gods, was that the Egyptian gods were not afraid to go to the afterlife. Clay pots that travelled during that time from ancient Egypt to ancient Greece, through travelers and the belongings of travelers who travelled between the two places, explained a Egyptian world of war, because the red clay pots were painted with black images of soldiers fighting a biblical war, and no one wanted to see the face of the beginning of co-operative labor or the ending of that era, and so the ancient Greek gods decided to wait until Ba' ll arrived on their land. On a land full of actors, the ancient Greeks only assumed that they would welcome fellow actors at some point in time in history, that *those who wear masks of animal heads during the day would one day arrive*, on their land. Athena, the ancient Greek goddess, the goddess on the roster of the young gods of their kind, the goddess of war and the patron of the arts. She, who was always in the know-how of whatever art came to her land, paid specific attention to the pottery that came from ancient Egypt to her islands. She wanted to attain something from ancient Egyptian folklore to call her own, because at that time that world was fighting a biblical war and running away from it, and her secret affection was always towards war and the belief in the necessity of war, so would it not have been a good decision for her to claim a signature symbol of a figure or animal or object from a piece of art that she owned, that came from a world that was facing the end of the world, while her world was still alive and running smoothly? However, Athena was inclined to the metaphysical and the spiritual aspects of life, and decided that the whispers that travelled up the Nile to her land about a owl she had never seen, -a owl that ancient Egypt honored so much for her ability to travel to and from the underworld, to unleash havoc against the enemy, a enemy that did not want to believe in a afterlife, that did not want to believe in the underworld, and that did not want to believe in hell,- would be a better ally and accomplice to her in case she would have to fight a war, especially because the owl was a symbol of wisdom all around the world. Athena found this to be more patriotic, to humbly choose a animal to become her ally, because the owl was fighting a war and Athena and the owl were both on the better side of the war, a war that would come to Athena's land as well. Athena did need a ally rather than a emblem at that time, so instead of pledging a allegiance to a symbol that she found in the art that came from ancient Egypt to her land, even though she found the artwork to be incomparable, she pledged allegiance to the owl because she needed a ally more. Athena was a patron of the arts as well as a god of war, and if she, at that mythological time, had decided to adopt some other animal as a accomplice or even a masked man, a masked soldier as a symbol, -like the ones she saw painted on those red clay pots-, perhaps in the future she would have been remembered better by the art world, in the modern day, as a artist. The owl who Athena chose as ally was a animal of war. The painted pictures on papyrus made by the ancient Egyptian gods, -within which were many images of the owl of the underworld-, did not make it to ancient Greece until the gods themselves left that land. No traveler from ancient Egypt to ancient Greece would be granted permission to take, -or could steal- such objects. Those paintings were well guarded, behind closed doors, inside palaces, because those papers would have deteriorated greatly by the change in environment from Egypt to Greece, ancient Greece having much more of a moist environment than the dry lands of Egypt, and so no god would let the images of the owl travel, but Athena banked on whispers running up the

Nile to her land which told her to adopt a owl for a ally even though she did not know what the owl looked like.

      The Nile runs from the south of Egypt to the north of Egypt into the Mediterrannean sea, to a wide open basin that is covered by water. The basin is underneath a dock existing there and spans quite far out and away from the most northern tip of the river. To dock where the Nile meets the Mediterranean is very treacherous. To drop your anchor there is dangerous, but there is only one port there, so everyone passing the basin always dropped their anchor there *to stop at a place where a river meets a sea*. The basin is a generating plant. The world can benefit from every infamous river that meets a sea, and for a river that wants to be remembered, that river can benefit from pouring out onto a basin. At that very spot where the river meets the sea, because the momentum of the river is shushed and the current of the sea is disturbed when a river meets a sea, if there is a basin there the momentum and shush can be used to generate electricity. Seafarers passing by that spot, where the Nile meets the Mediterrannean sea, will enjoy a whirl of wonder and the basin is underneath this moment, the basin is underneath the current, the only place where a bounty hunter would want to drop a anchor. That basin was the center of the world to two types of people, and the center of the Mediteranean sea, and if it was that basin where Charybdis *absorbed a great quantity of seawater, attracting in her whirlpool whatever floated by; three times a day she threw water back in the sea, and three times she reabsorbed it,* -<u>Myths, Tales of the Greek and Roman gods Lucia Impelluso</u>-, then surely that basin, that powerhouse, a natural resource was a location protected by the sirens of Greek mythology as well. Has it not been known that the sirens were phantoms, who could walk across water, and because they have always pledged their allegiance to protect the treasures of Medusa and somewhere underneath the ocean is Medusa's lair, and so surely they sing at the basin that is underneath where the river Nile meets the Mediteranean sea, when there is someone unworthy searching for treasure there. And if that basin is a place around which Odysseus must have sailed to return home after the Trojan War, and there he avoided shipwreck, if he was a pirate, he could not anchor there, and if he was a commandeer, he could not anchor there, because he is not a woman. If we steered north the Nile towards the Mediterranean sea and we could not drop a anchor whilst meeting the sea we would be on our way to Ancient Greece and we would have missed the spot on the map, if we were bounty hunters. But the ancient Egyptian gods were searching for something more than riches when they decided that they wanted to finally leave their land for certain, and forever, and so, even though they could have dropped their anchors just before they reached the basin, because it would have been convenient for them to do so, they avoided searching for treasure in a place where modern day women would search, because they were looking for the afterlife. So a river which was once a prominent natural resource to the ancient Egyptian gods, -beneficial to survival when farming depended on irrigation on their land,- became a mythological river, because the ancient Egyptian gods knew then that their art was too still for them to make it into a moving picture which is what a chart of life should be. So they left their land, to avoid the death of a art era, which was the ancient Egyptian art era, being, the underworld: they left their land to avoid death itself. And if the river Nile really was the Styx to the ancient Egyptian gods, they defined the river with a infamous name to celebrate mythology and Greek mythology. The Styx leads to the Mediterranean sea and to ancient Greece, literally, only according to a metaphorical transliteration between the mythologies and faiths of these two ancient traditions, but also on the face of a map. As for why the ancient Greek tradition favored this exchange, it is because though they could not use that river to travel downwards towards the underworld, the Greek gods thought that river would bring them gods they have never seen, from the underworld, and collectively they are called Ba 'll. This turn of events, which was when ancient Egypt arrived onto Greecian soil, which happened during a profitable age in art history, which was ancient Greek art, ensured for the ancient Greeks that they could have a future, during the renaissance, because the underworld was not a theme during the renaissance: ancient Greece enjoyed a renaissance when no Egyptian god was remembered. Ancient Greece expected the arrival of Ba'll and did not have to leave their land searching for a underworld and could instead make expensive artwork about their traditions and people at that time. Ancient Greece avoided the sirens who were singing them to shipwreck, which was the urge to travel from their islands, because during that time everyone wanted to know what the underworld was like, but this auspicious event, the arrival of Ba'll to their land, ensured that ancient Greece would know what hell was, and endow the future of art which celebrated their mythological age, with a concept of hell of which we have not seen, even yet, as benefitting more, any other era of art other than the renaissance, if the plight of a art era is to be remembered in the future by the modern day man and woman. *The concept of Hell benefitted us,* and we here in the western world have benefitted from the concept of hell painted during the renaissance.

And as for the face of the map, -which gave a reason to both these ancient traditions to feel helpless as the winds of change ushered in a compatible future between these two different types of people,- it ensured that no one out of these two types of people could go downwards, on the face of a map, to find the underworld. The reasons as to why what is down is up and what is up is down on earth is because Egypt is the underworld on a map of antiquity. The route to treasure, the route to making two worlds one, is a map, and though a map is no longer a work of art, there surely is one map in the world that is: a map from antiquity. The paintings we have done so far are all a response to the stories we have heard from the mythological age. The sculptures we have created are all a response to the stories we have heard from the mythological age. As if somehow the ancient Greek and the ancient Egyptian traditions were pious too, despite the fact, in the example of ancient Egypt, their economy eventually did collapse, and despite the fact that in the example of ancient Greece, for a long time they were forgotten and this is why we did experience a renaissance in art history. Because if everyone during that fateful age, then, actually did believe in Christos, then these two types of people would not have experienced a failure within their unique societies. But this one map, a map of the world from antiquity, -and perhaps it could have been tattered and fringed, if it was used to find gold, and if it was used by sailors to find gold, but it isn't, and instead it is cold plated,- it makes true the stories from the mythological era like how paintings make true the rest of history, because the entire mythological era was about making one of two worlds, the firmament above the firmament, and the firmament below the firmament. Sometimes I want to believe that it was not only the winds of change that caused this fortunate event to transpire, -which was when ancient Egypt and ancient Greece met in person-, and that these two ancient traditions did not follow blindly while taking part in fate to reach this goal, but I am Canadian and if the pieces of art from these two traditions are artefacts, then my heart may be less swollen for the artefacts that did not come from my land, but this fateful goal that was reached by these two traditions was what it was, so that the way could be led towards the future, so here, in our modern day, in Canada, we could uncover the past while being in the future like we are bounty hunters. *So that it wouldn't be known the gods are still alive*, and so we would have a past, and not a past alone, but one rooted in understanding a tragedy. And because the entire mythological era was about making one of two worlds, only a representation of our world, that is a sphere, that can be opened up and made flat, can be used to make two worlds one: one world is the earth, and one world is the map of the earth, one is the firmament above the firmament and one is the firmament below the firmament. The map of the earth is the firmament below the firmament. But the Egyptian gods knew they had to float up the river to get to the underworld, and up the river to get to the afterlife, so they had to float north, upwards, to get to their underworld, whether they were to drop a anchor under the current or not, and whether the underworld was under the ocean or not, and this is because their port is northbound, they face the basin and that is where the River Styx ends. Their upwards is down to ancient Greece. Only on a map did the ancient Egyptian gods actually steer upwards. Because if you sail the Nile you have to follow a map. Because if you are sailing the Nile you are a treasure hunter. And that was their dilemma because they always wanted to go downwards. The map became the most valuable asset. Only the gods from ancient Greece could ever go downwards to the underworld, downwards based on the realm of a map. Except when the word arrived that Athena, -a young god for who the Greecians built temples to worship,- would need a ally in the future because she is the god of war, and it was the ancient Egyptian owl of the underworld who flew through the whispers running along the Nile to Greece, the Greek gods stayed on their land because upon discovering that Athena was growing up they knew war was coming to their land. Word spread along the Mediterranean sea to Ancient Greece about the owl of the underworld, and that she met with Athena in a dream, and then both these worlds agreed that the owl is a symbol of wisdom, and to the ancient Greeks, a symbol of the old Minerva. And so the Greeks waited for those who wore masks of animals to arrive to their land, those who had spread the word of a owl of the underworld. In looking back there was no better river to be the Styx, for the Greek gods only wanted the Egyptian gods to land on Grecian shore, because it would be the arrival of the underworld, of Hades, of the unseen, of the gods who wore masks of animal heads during the day. There were really only two breeds of people at that time. The gods from above and the gods from below. They lived in the Fertile Crescent, they fashioned it, out of the eastern Mediterranean Sea, once the gods from below left their land to live with the gods from above. Oceania was the circle they fashioned of a territory of parts of Greece the main island, specifically Athens, most of the scattered islands that we today know to be territory of Greece, Crete and parts of Turkey, Cyprus and northern Egypt, and of course the water within that circle which is the Mediterranean Sea. That was Oceania. Islands and water. Oceania originally meant islands of the water, and that meant the islands that are supplied with fresh water mostly from the river Nile. We call the gap in the map, the space in the Pacific ocean, -when seen on a map of the world which makes the whole round world flat-, Oceania, because when the afterlife was reached by the coming together of the gods in ancient Greece, it was That Oceania that faced the sunlight for as long as the afterlife existed in the fertile crescent, and the earth stood still. The celebration that took place in Greece when Ba'll arrived was a resemblance of eternal life, the image of heaven as a everlasting existence in a place where the fire light never goes out, something on a ordinary day we live past and avoid continually because the sunlight always arrives to shine light on our day. During the night we manage to live in the afterlife every time it comes by, by pretending that the Oceania in the Pacific Ocean is the center of our world, and when that time comes about, Apollo thinks he has finally conquered motion until the world begins to turn again. Today, in tribute to the hidden qualities of gods, and that we can only always go forwards when we decide to travel and it is only on a map and of the directions of a map that the north, the south, the east, the west exist, we have two Oceanias. We have two Oceanias so that the frescoes in Greece of Poseidon can come alive during the night. So that we can dance with the dead there. So that the southern-most point of Greece becomes where we can one day dance with Hades, and so we can always remember the southern-most point of the Mediterranean is where the Nile meets that sea. So that those with liberal views, who will put aside logic and sense when searching for treasure because treasure to them is not treasure to us, when they find a bunch of maps hidden under the Pacific Ocean, they will not leave them there, they will bring them above the waters. They will read the maps like scripture but they are not scripture, for in our present day we no longer look back into the past for scripture, for what was written before the bible is not scripture. A map made up from the beginning of recorded history is still a map so long as we know one important thing, that it is not a artefact. We look at artefacts as beyond age really, before our coming here. Something we are not able to recreate because the hands of the artist who had once created it are gold to us, it is a incomparable and unachievable success. This is what we all want of ourselves and the things we create and leave behind after we die. But those maps Are artefacts, except once we pull them from the water they become found treasure. Much like the treasures that are recklessly found they will become tattered and fringed. We only have one burial site of our most sacred maps but the treasure hunter will have to search both Oceanias for this treasure because he does not know if he's looking for a artefact or a bunch of maps, because he will surely think that when the ancient Egyptian gods did leave their land with their valuable belongings, that they did listen to the sirens to shipwreck, and that they did have to fight a foe like Charybdis, and that no one could compete with that much adversity. And the reason why a bunch of maps remain a artefact only when left under the water is because if we ever placed things under the water to be found later, it would be heresy because still to this day we are not quite sure where the firmament below the firmament is, and leaving things in hell only means that the treasure hunter would only want to go back there to retrieve it later, and that is playing with fire: if we were confident enough to leave something valuable to us in a place like Hell we would be having dreams like Betty Boop in Red Hot Mamma, 1934. We would even go to Hell to retrieve our valuables because we love them so much. There are trespassing laws that are implemented by divine law itself, if we ever do want to go to hell to retrieve a lost object. You could even lose your life forever if you decided to enter Hell for the wrong reason. If we ever placed things above ground on soil in forests to be found it would be treason. We have a fertile full moon which consists of water and islands, there we can either be heretics or traitors, or both. When hunting in the Pacific Ocean for treasures we can only be heretics because what we drop into deep waters to be found later or not to be found at all is always something we assumed would make our household benevolent beyond the highest king so much so that we wouldn't even want to bury it on land. Truly this bunch of maps has already made us more benevolent for we have read them all, but to possess them again would make it unnecessary to travel. Most treasure hunters today wouldn't want to be either traitors or heretics but they cannot pull up a bunch of old maps to the surface world without being heretics, because though they didn't bury such peculiar treasure, they took part in finding it for us. The firmament below the firmament needn't stay there and up above here we have a map too, it's just not anything but the generic calendar.

    The future and the afterlife is awaiting us like one time it was waiting for the gods. The future and the afterlife are one and the same, but it will arrive one day on earth. Without the face of a clock we will never know how to navigate through a map in the modern day. A compass will never help us because though it is calibrated to detect the hemispheres we can never be sure on a adventure like this if a compass has brought us to the right place, because we don't know where that place should be. Depending on how north you actually go, then that compass will freeze. When you go that far north there is no more of a hemisphere but only a season. If we Were to use a compass to find this place on a map, because we had a bright idea that we could find a unknown location with a compass, Our bright idea is that once the compass points to every direction we must have reached Oz. But Oz is not a season. That place has no lights in the sky at night. It has no aurora and no borealis. That place has no snow that is as thick as the ice that has melted a little bit in your cold drink, which is now easier to crunch. That place has no snow that is soft right at the top of the sleet, right at the top of the snow cap where the wind blows trails of snow behind it. And there is no igloo there. If so, still, a compass cannot lead us to any season, -and there are four different seasons,- because a compass would still work if the earth stopped spinning. Our very clock is a compass we can use. The face of a clock has four succinct hemispheres and the clock only gives us two days of time, night and day, until it cycles again. That is a image of a globe and one revolution of this globe, of this earth. A compass was originally meant to be in the image of a globe, but because it was made to assist us while we use a map, it lost its genuine quality to be a circular object that has sensitivity for direction upon the earth because only on a map are there any directions. A compass will always only lead us forwards. It was the worst invention ever made. The season that a clock can lead us to is night, like the afterlife for the gods was a perpetual night, a perpetual winter. A clock only stops for two reasons. One, it's machinery is too tired to work, and so the arm stops revolving, and two if the earth stops revolving. It will be the clock that will tell us when the sun will rise again after the afterlife. It will be a frozen clock once we've reached it. The map was already used once as a compass to the afterlife. It is unusable now for us in this endeavor because we have no markers like a river to lead us anywhere, because the continent of Canada is much larger and spans a great distance, but we have a clock, something the gods never had. We are fools to look for a future in a clock. Only because since the creation of This devilish machine it has not stopped. But when the clock at Greenwich stops, when our pocket watches stop, whoever is there when we are there, whoever we are, and if we are in search of a place at that time, the most obvious guess is that we've reached it. The most obvious guess is that at that time the earth has come to stand still. Because not only do we need to be at the right place at the right time then, but we need to be in the right hemisphere and in the right season, and this time we will have electricity so if the season never changes again we will not have to live in a tragedy forever.

  Athena, a young Greek god, was about to face Minerva when two worlds were becoming one, when two breeds of people were going to meet each other for the first time. And when the gods from below and the gods from above were about to accentuate Heaven and never forget Hell, there was a concern which grew from ancient Greek philosophy, that another map of the world that described what the earth looked like during a ancient time would become popular and would come to reclaim Heaven and Hell. Would come to reclaim the firmament above the firmament and the firmament below the firmament. Would come to reclaim the past. It would be another origin story for the countries of the world. A idea that would compete with the earth itself to become a part of the chart of life along with the other celestial bodies, because a belief in a idea like this could claim that the earth does something worse than hustling and bustling. That idea was Pangea. The Athenian tragedy, the Dionysian celebrations, that took place at the acropolis, were about to end, -but only momentarily,- because no matter how attractively Pangea has been rendered on a globe of the earth ruins would be ruined if we could not stop the tectonic plates from shifting. Like the belief in the stillness of a green-bronze sculpture of Minerva, we were destined to avoid a natural disaster and many of them, when two world were becoming one. Natural disasters who would take away our oceans and seas and would cause sculptures to topple, and would take away what was once a missed opportunity which is to watch those sculptures come alive, during a celebration. The ancient Egyptian gods arrived to Crete, a island of ancient Greece and Athena faced the glorification of the fresco. And so the patron of the arts received that one failure. That a Dionysian celebration and the arrival of the afterlife would not happen at the Acropolis, for that one night, but Athena could bank on a whisper running upwards towards Greece from Egypt along the Nile river of a owl that would solidify her identity as a Greek god and patron of the arts. *Athena at that time would have to introduce tremors and shaking- as a result of lightening hitting the earth, as a result of the tectonic plates moving-, into her Dionysian celebrations, if she wanted to be the leader of any cult,* which means she needed to learn how to dance. Ancient Egypt arrived to the shores of Crete and it was on that island where there would be the first Dionysian celebration, it was the glorification of the fresco but during that moment in time, the sculpture of Minerva would not be forgotten even though Athena was, *it may be that Athena, the young Greek god, would have to assume a duty to war before art*. Long before ancient Egypt arrived to the shores of Crete, ancient Egypt glorified a black owl through art, who could go to the underworld and come back. That black owl was a iconographic image of wisdom for those people- like how the owl is a symbol of wisdom for our people,- a beautiful animal that became a symbol because of the industriousness of those people and their great will to make beautiful art. It was because they made wood-block prints in black ink of the black owl, that the symbol of this bird and the meaning of this symbol remained true to the character traits we believe this species to have in the western world like wisdom: *this was a western art iconography*. Black was a color used to decorate the masks and golden sculptures and pottery and funerary objects in ancient Egypt: it was a royal colour amongst the ancient Egyptians, but the black owl, in ancient Egypt was a bird with a soul and avoided becoming part of any mask the ancient Egyptian gods wore, and avoided becoming a sculpture, -whether those gods wanted to depict the owl in pure gold,- and instead became the comedic and awesome authority to funerary rites in the form of a story teller about the underworld. Sometimes the black owl was depicted as flying, or as perched, both in the pictoral stories of the ancient Egyptian underworld and in hieroglyph, -when incised, when sculpted, when painted-, but when the black owl was painted in black, printed, it was then, it was that symbol that became a recognizable and sincere and stark and noticeable motion, and successful endeavor achieved for the black owl and Egyptian art. The owl of the Ancient Egyptian underworld was black. The animal, the owl, was a artistic, cultural exchange between one Greek god and the Egyptian gods which happened when two worlds were becoming one, as if the owl was one of the first iconic exchanges, a popular artistic exchange, because the owl of the ancient Egyptian underworld was black, black like ink. The owl of the ancient Egyptian underworld could be printed, a wood block print, in black. This owl of the ancient Egyptian underworld could be painted in black, and this black owl could be a hieroglyph: the owl of the ancient Egyptian underworld was copywritten, no one would be able to tell the story of a black owl better than the ancient Egyptian gods. Athena was industrious. She wanted to make art. If the black owl came to Athena in her dreams, which she did, it was a snowy owl that came to her, one day, in her waking life, and this owl, who in spirit, was much like the ancient Greek goddess Minerva, cared more about war than art, and if this owl, who Athena did name Minerva, did care about art, she often concealed this from Athena because what was in the best interest of Athena was to know how to wage war at that time: Athena made a ally of a brown and white, snowy, owl. The future for both the gods from below, and the gods from above took place on the firmament above the firmament, as seen on a map from antiquity. The black owl became a missing person in the future, someone everyone was looking for because she was a piece of wisdom from the past, when the gods from below arrived on another fertile land. In ancient

Egypt the black owl was ally to no one, a snowy white and brown owl became a ally to Athena because she lived on the firmament above the firmament, *when regarding a map of antiquity*. The female owl was white like the heavens, and white like snow, and brown like wood. And though the black owl did come to Athena in her dreams, the type of war that Athena never knew that she would see, upon snowy mountain tops, upon landscapes that remind us of where Norse gods would war, was a place the black owl did not want to war, because in a war like that, if one failed there would not be a future. In a war like that there was nothing above the mountaintops and nothing preventing you from falling. The black owl did not represent death, unless that owl represented rebirth as well, and on the face of a snowy white mountaintop, in the face of a war like this, losing is a concern, and there is no afterlife and there is no re-birth there if one loses. Athena, also being a patron of the arts, and who did want to make art, decided that a black owl was beyond her realm of expertise to become companion with, to make art with, and to make war with, especially because a black owl never flew to her in person, as if the Egyptian underworld foretold the future that perhaps Athena would not see a war, and if she did there would be snow where she warred, and her war would take place during the day, and so the snowy owl came to fulfill a companionship with her. During that period of time Ancient Greece gained a owl and ally to one of their most favored gods, a god of the future for those people, and ancient Egyptian gained a incomparable concept of the afterlife and direction to it through the whispers in the wind from ancient Greece, which caught the downwards current of the Mediterranean Sea towards Egypt. In between these two worlds the owl maintained a cultural theme: mysteries of the afterlife which were pristine, white, snow for both of these cultures. The hieroglyphic of the owl is a painting when it is painted, a drawing when it is drawn, and a relief when it is incised or built up, but it was the first rendering of a representation of wisdom we had ever seen. It was a representation of wisdom and this was exemplified by the fact that Athena, the patron of the arts, wanted and needed a owl to be her ally. The Egyptian gods never had a owl headed god in their circle. They had a jackel, they had a eagle, but not a owl, though the owl was the only bird in their hieroglyphs and artwork when the story was about the underworld. We, like the gods, worship celestial events. We are in awe in how the world turns, and how the animals live, but attributes we want to attain, we learn, that is the root of wisdom. The Egyptian gods did wear masks of animal heads and rendered themselves masked on thrones. They celebrated like the darkest Dionysian parties, but one event they were lacking was a exodus, something that would take them to the afterlife so that they could live forever. But what they didn't realize in the beginning was that they needed a tragedy, they needed to become a part of a play, because the ancient Egyptian gods did not find death tragic, every single Egyptian god needed to be in that play. A play where one actor believes he is already dead. That he has died and lives in the other half of life, the afterlife, the underworld, post, because the Egyptian gods only believed in immortality after death, after a season: the art era the Egyptian underworld was filled with so much glory, and gold, that no art era from their region of the world could succeed it, and the death of a art era, with no direction to the future, means the death of a society. The ancient Egyptian gods only believed in immortality after death, after a season, and yes that time is after one of the four seasons, the time is winter after fall. Ancient Egypt was once industrious. Farming was a second nature to those people. Though they did not experience every season, like we do here in Canada, - we experience summer, fall, winter and spring here in Canada,- they did know that they wanted to. Immortality was winter forever, for as long as until the season changed, and the fall was death. The underworld was a art historical period for the ancient Egyptians, the idea behind post-modern art, because we believed modern-art would not last forever, so we came up with post. The idea behind post was the afterlife to modern art but the reason why it didn't last forever is because no one wants to live in a tragedy forever except for the Egyptian gods. The ancient Egyptian gods would live forever in a perpetual night. The Greek gods found that tragic. We have to remember that the Greek gods also found a comedy in that very tragedy, because they knew that a sunshine filled day was a inevitable and that is why they enjoyed a renaissance which began in the 1400s and lasted two centuries. Because the one who thinks he is dead is not dead: the Ancient Greek gods never understood what it meant to exist inside of a tragic play, for realsies, when only one actor dies, because they only had themselves to be actors, and they were all actors, so they too wanted to discover what it felt like to live a tragedy. It was the Ancient Egyptian gods who prescribed the afterlife to the Greek gods, as if the ancient Egyptians were highly qualified in the medical sciences, which they were, and the ancient Greeks were too, a quality in each of these two breeds of people which could have caused them to congeal their appreciation for art and the muse, and the pose, and the male and female body, and sculptures of figures. But why winter was the afterlife to the Egyptian gods was simply because it was the only season they never experienced, and that was a tragedy to them, as they were a part of the western world. Because death itself is not a tragedy, as long as it is certain that every actor will one day die. The death of one in youth as opposed to old age is not a tragedy, despite our common day view, because we were meant to be immortal, the death of a god is a tragedy because that is a death of a ancestor. But since the birth of the tragedy we've lived in one, and the changing of the season has not rid us of it, because since the birth of a tragedy we've ensured that every tragic play is performed in the winter and we forego the fall, every theatrical season, and so we begin our theatrical season with a tragedy rather than a comedy in every reputable playhouse,- like for example Christ carrying the cross is celebrated in winter- but if we began our theatrical season with a comedy, we would be performing our stories in accordance to the season. What does winter and night have in common? It is in fall when the days are the shortest and the nights are the longest, just when winter begins do the days begin to grow longer again, at winter solstice,- and that is a tragedy for some-, but that is a common winter day, and we enjoy that day too. But in winter the location, the longer the darkness pervaded...the longer the night pervades, the cold sets for a eternity, *let us take one gaping awe at the globe of the world*. This is why we have two days, one where there is sunlight and one where there is night, the sunlight arrives after every night so that we will not live in a afterlife, so that the cold will not sink in deep enough to sustain itself, because in some places in the world it does. For every hot day the night is colder, for every cold day the night is colder, but to live in a perpetual night, though that night is cold, it won't get any colder because that is the common temperature of the season, the season being night. Truly the afterlife is a location, and if you are there, it will be winter, and the season there is the night. But the winter in the afterlife is a season too, because there is tundra there, because that winter faces a night sky within which is centered Orion.

Our constellations are arrangements of stars that to us shape out and embody mythological stories from the Western world, of figures and animals, and also objects. If those arrangements happened by accident it is a miracle, but the constellations were not here before our myths, they were here at the same time we first were. We are immortal only because we have skin. Our myths and constellations do not have skin: we are born and we can always choose to ignore the popular myth, but, the odds are that there is always going to be someone in this world that will remind us of those myths, to ensure that we have stories from the past to believe in. We want our constellations to have age, have a history, to prove that we have a celestial sphere that will not change in accordance to the arrival of modernity or any new idea, and so we've created myths that resonate with the stars in the sky. Nietzsche said that who is to know if the stars are in the sky or are in our minds, and he also demanded us to ask why we do not search for the untruth, these two questions are the reasons why every myth about our constellations must be true, and must be stories as well as mythological in the sense that they are old, unless we want a procession of constellations in our minds and nowhere to be found in our skies. For a man there are only two places the stars can be and we cannot have them in our minds if we do not welcome death along with them, into our minds, because stars die and are born.

Already today we have seen the birth and the death of the sun, and we can credit modern art for the death of the sun, as seen in Monet's sunsets. *During the era of modern art which is when we are living right now,* we have fallen susceptible to the hues of art history, more than the visible light spectrum, and we do want to view those paintings at night, with candlelight, and with the stars shining during that night, outside. And it was specifically within the years of the era of modern art we as artists collaborated with daily life, during the creation of the grid, and finalized, that, fine art will never be about anything else other than the great myths. We remember every myth inside our minds, our mind is the archive, and we reproduce the great myths in art in the subliminal search for treasure. Sometimes it seems obvious that we should desire decadence and stop painting, because we think we have a lot of good paintings, if only the hues of our paint, the smoothness of the stone was not decadent. It is in every painting up until the death of modern art that we see the birth and the death of a star, that we see the rising and setting of a sun, and we see the birth of a moon and the death of a moon: it must be certain now that during the era of modern art we made new myths. Modern art is the last place we can use to put together myths. Modern art is the end of paintings that since the beginning of our acknowledgment were painted centered around myth, and this era is slowly building, and we will see new myths, and we will all decide if they will belong in history because most of these myths have nothing to do with the stars: modern art is a bold night sky. If the stories that arrive from the era of modern art, our present era, do have anything to do with the stars, surely some of those stories will be about a celestial language we have not yet encountered.

The last place you want to be is awaiting a new season.

There is good reason to believe that the nebula within Orion's lower torso gave birth to our procession of constellations. What this means is not only did that matrix give birth to the stars in Orion but it gave birth to all of the stars within our celestial sphere if not all of the stars in the universe.

The twin of every star dies inside our earth before it is born in our night sky: this reminds me of The Seven Deadly Sins and the Four Last Things by Hieronymous Bosch, because in that painting Jesus himself is in the center of the earth, putting a end to a sin, from the center of the world. The twin of every star dies inside our earth before it is born in our night sky. Who's to say that this is not how our constellations have stayed the same for centuries. This ensures that a star has a mind it dies in, which is the earth, and so that the mind of the earth can continue to die as it begins to awaken, so that it will never awaken and never become thinking instead of feeling, for if the earth told us everything, everything that was wrong as well as right about the world we would be too afraid to hear this answer from the geyser.

The stars die one by one in our earth before they are born because anything that is born above the earth's curved surface, with having no roots in the soil, must fall first, like the religious man. Orion drops a star unto earth from her womb, because like every mother she too discards a degenerate egg before she gives birth to the related child. She drops that dead egg onto the earth. The corpus luteum. A completely dead star. It is as small as the corpus luteum of the body of a woman, and it falls onto the earth as rust. It is sulphur that falls from her celestial womb, onto the earth to continue to fall through the cracks and water of the earth, and quiets the mind of someone who could be a mother. The dead egg that comes out of Orion's belly is a unborn child, is twin to the one that will be born in the night sky so that the one that is born did die before it's death, and the child's deathbed lay underneath it, light years underneath it. This is a universal metaphor for oogenesis, for a oogenesis that was meant to produce two children in one ovum: twins are the gemini of a mother. It is highly likely that a universal mother, who is Orion, would give birth to the same constellation again, to live in our night sky, and so her menstrual cycle will ensure that she will not. In the universe, in our night sky and inside our celestial sphere, there is a competition amongst the most virile constellations. Orion has been the home of a profitable nebulae. But she is willing to forfeit one day, and Uro the bull does want to own all of the future. She does want to be a home to a womb that could give birth to a future. Gemini, the bathers, two beautiful women are a better fit to give birth to a future that is not that far away, but if that was so then that would mean that the female bull could not compete with someone, where as before she could compete with anyone, and now she would be competing with two women. If it wasn't for sexual differentiation, the urgency of locating the gender of a child, the first fork of oogenesis, to rust would not mean, also, to bleed and we would not have a affinity to starlight: stars are born to be either male or female. Orion takes from the only unwilling man within our stars, the sun, and this is how the sun is young and old. And this is when the sun is not Apollo. This is when the sun is just a sun in space, and is a star in space, part of no constellation. This is when the sun is a celestial body and has a magnetic field. Though Orion, though she is not our oldest ancestor, she is our oldest ancestor who's body is shaped out of the stars of our sky. Her body is the quintessence of the female form, the shape of a mother who gave birth to a sequence, towards a bloodline. If a star would fall, it would fall with the knowledge of a celestial sphere we have never seen and what we most probably don't want to see, causing rabble. If life was anything like Hieronymous Bosch's painting, a circular chart of ramifications and repercussions, Jesus would be the undertaker of the twin of every star, and we would surely only have one celestial sphere.

And this Jesus is not a rabble rouser, he does not want to raise a little hell, and if he can raise the dead, he would do that like a magician would. If we want to make the origin of our stars and star constellations simple, and if we want to have order within our celestial sphere, and we want to not care if there is disorder in the universe, that generator cradled in Orion's womb gave birth to all the stars. *If the stars in our night sky can equate with the myth we embody, then only one nebulae can give birth to all the stars*, -and then there is a universal mother- because the birthing cycle of just one nebulae is much more competitive than the birthing cycles we do see competing for space, food and life in the other kingdoms. But when we are talking about the birth of stars, obviously, if we believe in one static celestial sphere, when we are talking about the birth of stars we are talking about the past, but that does not mean that a star is not a species and existent to be charted in our map of life, the star will be one of the few fearless entrees. Our greatest accomplishments as mankind is having observed a hierarchal system of species and having created the electrical grid system in Canada.

It was the beginning, the Scala Naturae was a biological classification system that made even the rocks and gravel and pebbles alive, that made the ocean and the tides alive, the moon alive, the sun alive, and it made gravity secondary and in accordance to our will, *for a star is one of our species, and gravity cannot hold a star down.*

The wind and air will make it into our chart of life. At one point in time we realized that we needed to keep order in our world, on the earth, in our earthly atmosphere and in our celestial sphere, we had to make the choice of if we wanted order in the universe or within our view and reach. Right now there is no order in the universe in places we cannot see, maybe because we do want to see a shooting star in our night sky, and we chose it this way so we could include our night sky -and day sky- in our biological classification system. We have two stars in our biological classification system, the sun, and what is a star. Because the sun is named we cannot exclude it from the chart of life, and we would not want to, even though we cannot include our sun in our celestial sphere amongst all the other stars, in a chart of life, because we cannot see the sun at night, and we never will. Our sun is immortal, in this chart alone, because it is named *the sun*. Another sun may want to enter the chart of *life, but it won't*, because in this chart of life, the sun is not a star, and if the sun is a star, it is the only star that shines during the day. The sun must prove it to us today that he is in fact Apollo, because the result of the belief in the sun god Apollo, in the modern day, as a result of more than centuries of believing in this story, will result in many smaller suns who will all claim they are Apollo, -suns derived from unclear myths-, and suns that are not perfectly spherical, like our sun is, as if every morning a new sun is born and rises, and dies when it sets, as if a shooting star or a comet passes us by every day, to bring light to a new day every day. Perhaps, when we wake up in the morning to see the sun rise, and we wait all day to see that same sun set, and we wait during the night until the next morning to see the sun rise again, it is the same sun that rises again: in the modern day we may have wanted to see a entirely new sun rise, and maybe we might have wanted to see that every day. Why is it in the modern day we are not certain we want our sun to have age? Maybe because we do not want to call the sun Apollo. The dead moon, the blue moon, the white moon, the yellow moon, the orange moon, the red moon, makes every singer of the song we sing about the moon bubbling with joy towards the moon, and unwieldy contempt for anyone who could say that we could live without a sun. But the contempt of a songwriter towards the sun, who wrote a song about the moon, comes from his very own anger that he cannot prove to the world that it is more valuable to have a night time where we cannot see the sun. The moon is like the infancy of a star preserved, yet the moon is such a old relic. The moon is a drama on its own of what it means to be immortal. Because of our moon we remember our sun: the sun's greatest adversary is the moon but he is also his greatest ally.

   Should our earth have a molten core or be hollow? The lava that bursts out of old volcanoes proves certainly that there is a deep reservoir of hot moving substance somewhere within our earth especially because lava comes out under our oceans too. The shifting tectonic plates sometimes causes this as the earth rips open. But the very center of the earth must be more like a moon, a small perfectly spherical core with a definite crust, only simply because sound that reverberates in our environment could not possibly be heard if our earth had a molten core, instead of a core made out of quartz. Would it be odd to think we have a moon in our earth? Would we be surprised, maybe happy? Or would we shudder at the idea that our entire earthly existence is built on a dead star, *if only every moon was like Dionysus*. If it is a dead star inside our earth, then certainly it would be the first star ever born out of the womb of Lady Orion, the closest light year away from her, the closest distance in kilometers away from her, the oldest star in the universe. Perhaps the earth is the oldest planet built atop a moon like a house built on the side of mountains. The sun, though is a star in every way, but because the sun lights up what is called a day in one revolution of the earth, is the counterpart to the diorama we have of the universe. The sun is one celestial body that we cannot see at night, which we are already certain of, but that we also cannot see during the dark. This makes the sun different from other stars. If we shone the light on the diorama of the universe during the night, which is a season, we would not see the sun, we would only see the sun if we shone the light on the diorama of the universe during the day. Though there is always light cast upon one side of the earth from the sun, the day is not a season, the sun unfortunately cannot be seen during a season, during a time of celebration. So, there is a day every half of the revolution of the earth but when we did shine light on our diorama to see the sun, we shone light onto our diorama from the right, because we are conservatives, so if the light of a day did not reflect onto the sun we wouldn't even see the sun during the day in our little universe. The light of the day can only be of use of a sun that wants to be seen, if that day is of the central Western world, because when we turn on the light to our little diorama, without touching the globe of the earth we are finding self-satisfaction that the day is progressing in the West which would mean that the sun does not rise in the east.

One star became a moon. And supernovas and blackholes are not the last stage of the life cycle of all stars, the last stage of the life cycle of some stars is a moon. Fourier's speculations of how our earth will one day be center to four moons could not have been science fiction or a joke, could it have been? With as many star's deaths the earth has harbored, the earth, our reservoir for sulphur, why would Fourier speculate that four moons would come to circle us? Because the earth has been the deathbed for all the stars. Perhaps only three more stars will die, -who currently exist in our universe,- will reach the state of what is a moon and every other star will reach their full magnitudes and from there will stop growing, never reaching the end of their lives, never being born again but yet not dying: is this not the moment, was that not the moment that we gained a celestial sphere? Maybe these three stars want to leave the universe, to be center to our celestial sphere, with the earth, and with the moon. Other planets have moons. Some have many moons, or so we call their satellites moons, but in our myths there was always only one moon. There was never a mention of Europa or Ganymede or Io existing as moons of some foreign planet but these are the types of myths that men can build a phylogenetic tree off of: outer space is male. Why is almost every moon and planet and star male? Men can string along the moons of this universe to make comparisons- with themselves- on their phylogenetic tree, they can use the names of distant moons from distant planets: if a man wants to find his taxonomic origin, everything that is male can be on his chart, everything that is male can be in a chart that explains his primal beginning, a chart that has a past before man arrived to stand above the earth's curved platform, a past which is the residue of the male gender. So the moons of distant planets could be on that chart too because outer space is male. Phylogenetic trees are beneficial when we find a need to compare the growth over time, and origin story, of any existent. It is traditional that every living thing on a chart like that should be the same gender. We have for some time now used carbon dating to prove the birth date of species who existed a very long time ago, and we used burial sites: by digging deep into the soil we found the origin story of species, but we may find today that a philosophical quest to chart the origin of man will cause moons far away to be noticed because they are male. A woman has a strong phylogenetic tree because who is a woman has many identities due to our current modern understanding of what a woman is and what is female, and our drama on who is a woman and who is a young girl. A phylogenetic tree where the origin of a woman is found will be filled with many ancestors to a woman, and some of those women will represent archetypes: on a woman's phylogenetic tree there will be the hero, the witch, the queen, the mother. As a result of a hard-won history of a woman is that the archetypes that are ancestors for her are often always good role models. Often we find that men do not want to be like a ancestor. Often we find that each man would rather be individual and alone, and so on a phylogenetic tree where man exists, we will find less archetypes and we will find distant moons. *And if we titled a tree of phylogeny, family, then a man would fit there too.* But for a man to compare himself to himself, or other aspects of himself, he can use the lines that extend from the universal nature and he can do that much easier than woman can, because moons and earths are male, mostly, and these existents serve as family for a man, like a blooming rose can serve as family for a woman on a phylogenetic tree. A blooming rose can serve as what is feminine or as a symbol of femininity for what is female in the family tree of a woman, in a phylogenetic tree of a woman, and a little black cat named Rak says: *I can be on your phylogenetic tree as well because I am female.* I am not the salesman to a chart like this, not to a chart of life or a phylogenetic tree. It's pretty certain that the only person who can sell you this chart is the famous bull Uro, and Coco Chanel. If there is anyone out there that is looking for beauty or ancestry or something to be in awe of, a tree that looks like a fork, like a fork in the road, is a mesmerizing mystery to look into, like a phylogenetic tree, like oogenesis, like the tree in the yard whose trunk grows into a fork. But so long as gender is the priority in our phylogenetic tree, then we have a tree.

  Fourier, with his prophecy of four moons and a ocean of lemon juice: as much as it could sound like a political satire for the upcoming globalization which is nearer now than ever before, it seems more so that he was devising a excuse away from the seasickness that we all feel when we sail over morality, -which was what Nietzsche said we should feel when we sail over morality-, that four moons could harbor the weight of a heavy soul. *To make a map of the earth*, so that we can know what outer space looks like, so that earth can extrapolate and a budding economy can grow, as licentious as it is, but in the face of globalization, in the face of a population overgrowth which we could not control, we still must monopolize on this earth. This map must suggest that the four corners of this earth all extend from the equator. This will glorify the ever-present importance of the kilometer as a measurement of use to measure distances between celestial bodies, and thus will revive the old myths and the mythical names of planets: if our moon is kilometers away from us, it can never be in a biological classification system of all life, if the other celestial bodies are also not kilometers away. What the moral question should be, which we sail over, is also what a feminist should question. How beautiful do we want our night sky to be? We can only ask these types of questions to the universe with love in our hearts because we surely don't want the tides to change. This is all about proving how beautiful the night time truly is.

      As for the moons to come who will circle our earth. If they are to be moons of our earth, they had a life as a star before they became our moons, like our moon was once a star. At some time in their life they shined with a maximum magnitude very similar to how our moon once did shine. Two more stars will travel back towards earth. Each one becoming the sun as the sun becomes a moon, but our moon never traded places with our sun. We can live without a sun forever if we have four moons to circle us. With four moons the seasickness that every woman feels will dissipate. It is a woman who battles the duty to overcome what is morality for herself, more than any man. It is a woman who competes with the pedigree of decreasing chance always and inevitably, which is oogenesis, and then, another inevitable divisive, indecisive loss which is that of the corpus luteum every month. Because of oogenesis woman is never free to expound her sexual maturity. The fear of a never-ending pregnancy should always be near to a woman, but will these four moons harbor the weight of a heavy soul for a man? In the end it will be a man who decides if it is logical for us as a society to pledge to a future of consumption just like it was because of the idealistic vision of a man that modern art died. In the modern day the modern-day man found no need to idolize a woman. He found no need to make a sculpture of her which was true to the form and function of her body. The last image of a woman, made by a man, was not realistic and a man thought that it was vain to need to create a sculpture of her which was realistic because he did not want to compare her to the infamous archetypes that pre-date her. He wanted to celebrate her in his own unique way, and then modern art died. Maybe that is why a woman needs a afterlife. Woman is surely looking for another phylogenetic tree to pledge allegiance to, -like Uro has a line of paintings,- so she can exist outside of the duty to the female sex and become masculine in the database of art history, like Lady Orion is masculine in our constellations and myths. Woman surely is looking for a alternate ancestry so at least she can prioritize her life and for short periods of time know that oogenesis is not the be all and the end all to the female body. Oogenesis is the backbone-of-a-tree that follows along with every woman, something every woman has in common. Oogenesis is the fork in the road that can immediately cause the world of one woman to change, which is the inevitable birth of a child, but sometimes women do want to change this inevitability too, even if they know they cannot. Oogenesis makes a woman unique, why would any woman be running away from that? A woman born in the modern day is forced to embrace technology because every man does. She has to compete with men, as if she is a man, to make a living, yes, but what is more of a challenge for her is to embrace technology, and with that, modern day medicine, to prove her existence to herself in the modern world. It is logical that we perceive contraception as the most modern medicine because it is a competitor to the oldest rhythmic pattern in the world which is a planned pregnancy. That we can prevent a child from being born by the use of medicine should mean to us that we have reached the end of a era. A woman born in the modern world has to face the reality that she can challenge oogenesis even though in a world where a man is powerful, and the man she loves is powerful, the man she loves may not want her to challenge oogenesis. Though we as women may hate to carry along this back-bone, this fork in the road, we as women still want to challenge the status quo, and now that seducing oogenesis and reducing it by our own will is something that is popular and possible by modern day medicine, well that is the status quo that we want to refute because it is modern day medicine that has the agenda today. A man who does not want the woman of his dreams to take birth control will often win on achieving this desire, because a woman will always want to challenge the status quo. Today it is popular to use birth control, and it is common practice for a woman to use birth control, and so a woman will challenge that: at the end of a era a man will still succeed over the desire of a woman because a woman will always want to compete with the status quo. Also, though, a woman may have ancestral mothers who do not want her to seduce oogenesis either, but they may not speak about it with a woman because that bold action may make a woman less successful in the world.

    And if the woman is depressed, a kind of depression that may always be near to us, to all of us, to the both sexes, because we cannot truly go beyond our natural perspective. If the woman is diseased with feelings of depression, and so the ailment is of the mind, she will be competing with two types of degenerate eggs. She is always competing with a pedigree of decreasing chance which is oogenesis, and with a inevitable garbage which is the corpus luteum, but if she is mentally ill, she will most likely be on a drug called lithium or a drug called divalproex. These two mood stabilizers are two degenerate eggs, which means for the depressed woman they each create a pedigree of decreasing chance, a hypnosis away from the inevitable, which is seasickness during war. There are two degenerate eggs. One is divalproex, and the other is lithium, two mood stabilizers. These two different pills are degenerate because they are similar to twins. In the world of modern-day medicine, because these two medications have too similar qualities and the result of their consumption should be the same, in the body of a woman, these two degenerate eggs are not sellable and so that is why they are degenerate. These two pills are degenerate because they are similar to twins, not unlike the possible product of oogenesis, a result which would be born, always, if there was not a fork in a woman's backbone, -the phylogenetic tree,- if sexual differentiation wasn't a inevitability. When twins are born it is a blessing. When the primary oocyte is born, uneven, it is a blessing. When a primary oocyte is born uneven and the reason for that is not only because the cell has a feminine side, but because the wrong child is to be born, because one feminine cell has two north poles and two south poles like a flattened map of the world where the pacific ocean is a gap in the map, then no longer is sexual differentiation a inevitability for one child alone but for two. If a wrong child is to be born, that is a inevitability, and that child will probably die and that is not a worry for any of us, but the oogenesis, that tree that existed to create that ovum can be inherited, it can show up in the life of that woman again, and this time she will not lose the child! In a predictable world, when we know that a corpus luteum can extinguish any wrong child, we should know that a oogenesis can never be blamed for the loss of a child. It is better to blame the embryonic sac. It is better to blame the umbilical cord. Then we know that if a woman does lose a child, she has lost nothing genetic.

   When twins are to be born, the cell must have a north pole and a south pole, and so if the popularized myth of the Gemini, being a twin boy

and girl- which has been popularized by the infamous Western zodiac- is true, then the map of the constellations that make mark on that one cell will make a mark on a moon that we have made a map out of, flattened, like a map of the earth from antiquity, and the result is meiosis.
Sometimes the constellation Gemini, high up in our night sky, appears to be twin boys, and sometimes the constellation, high up in our night sky, appears to be twin girls. When you read your horoscope in the Western world you will find that each writer of a horoscope may believe something different about the constellation Gemini: that the constellation Gemini is a story about twin girls, or a story about twin boys, or a story about a twin girl and boy. A oogenesis that tries very hard to prevent the birth of twins fathoms the birth of twins, only if the to-be egg is a moon. That moon is center to the celestial sphere that our earth is center to, but during the birth of twins, that oogenesis looks like a map of the moon flattened like a map of the earth from antiquity, that map of the moon is eclipsed by the map of the earth from antiquity because the map of the earth was made first. There is a north and a south pole to the moon, and to the moon in the womb of a woman. But because the moon in our sky has no territories marked on it's map, there is only one north pole and one south pole on the map of the moon, even if it is made in the tradition of a flattened map of the earth from antiquity. And so when gender is assigned for a twin boy and girl, at the primary fork of oogenesis, it is easily assigned because the to-be egg cell is a moon: the girl will remain on the feminine side of the cell which is the north pole and the boy will remain on the masculine side of the cell which is the south pole.

  As for lithium and divalproex, these two degenerate eggs are not unlike two jagged suns, suns that are of a oval shape, but when you have to take them, if you have to, if you are a woman rest assured they will most definitely not effect the uterine lining. It is never the status quo, -the one that a woman must always compete against-, to consume psychiatric medication, because it is always that less of the population of women, at any given time in history, will take that medication, and that medication does not change a commonality that women share. A man may not ever want the woman of his dreams to not take psychiatric medication if she does feel depressed. At the end of a era a woman may feel less seasick but lithium and divalproex are not the most modern medicine, so a woman will never refuse to take those two medications if they are prescribed, even if she is not depressed she will take that medication if she is prescribed it. But those two pills are like jagged suns, like a sun that would want to make it into a chart of life, even though he is not Apollo, which means in the future, if we cannot agree that in the modern day our sun is Apollo, we may find another drug that can help to ease seasickness in a woman, and that drug will be of the same quality as these two degenerate eggs.

    Either it is that we already know that our night sky is more beautiful than one shining star in the midst of daylight, so we are not easily enticed by the beauty of the sun, and we know that the sunset is because of the night time and the sunrise too. Why some of us enjoy the misty trail of night time better than the day may be due to us wanting to hide our face away from industry, because industry is not a companion of the farmer, and neither is modern day medicine a companion of the farmer, because modern day medicine should not become our husband or our wife, and after all our concern here is seasickness and how much we want it to go away. Would anyone find it more beautiful to see a new moon in our sky, and then two more new moons? We are not easily enticed, and most certainly not, because each time that happens there will be a new calendar and we still have not learnt the importance of the primary one: the farmer's almanac. If new moons were to arrive to make the earth center to their existence, the purpose of them would be to knock the sun off its spot as if we're playing pool with the planets, but truly, and really, if we can even fathom four moons to circle our earth, today, it is a modern day sign that we are playing a gambling game with industry all because we need more night time hours. If we stay awake at night and avoid the day, because we are taboo, because the day is too busy, because we want to see the night sky, the working hours of the day have gone and passed and the only legitimate attempt we can make to one day seduce the day, during daylight hours, is to wage war: the only preparations that are made at night for the next day are when we are waging war and this is the birth of tragedy because we cannot enjoy the day. There is another priority we take part in every night and that is managing the mitsein: a preparation for a wedding. This is what keeps the night time romantic and within reason, this is the birth of tragedy too, because no woman wants to pledge allegiance to two wars at the same time, but she will.

If you prefer to stay awake during the night you are a Dionysian player, but during a time of war we all stay awake at night, so the fact that anyone is mentioning Fourier and his four moons must mean now is the time of war. The mitsein, that crucial angle between a man and a woman who are in love, will exist forever, that war will subdue the day as much as the things that make it a taboo appreciate the night. There is no taboo during the day unless you have a forest in your backyard where the trees provide enough darkness and roses bloom in it. Daylight exposes everything about the person, about the plot of soil you till. No one wants to be seen while they sail over morality: if you are tilling the soil for the first time in a long time, you are sailing over morality. No one wants to be seen while they sail over morality and the ubermanche will have a plot of soil to till too. If you have a place of your own, because sometimes a room will just not do, then all those who are dying to enjoy the day, those who are just too much of a taboo, then they can enjoy the day too. There is always someone who feels the seasickness more. In a world of rank, of creed, of class, of sex, there is always one person ahead of the game, more tortured for that one moment they sail over morality. These are the building blocks of a family and a family tree: rank, creed, class and sex. Within a family tree, one by one we have sailed over morality, but we don't step on top of each other to gain such a title as the ubermanche. If we've all found our plot of land, the seasickness will dissipate.

There is a concern if Apollo himself is not the sun of our day, of our every day. Apollo is the sun we wage war against. There will no longer be a clan competition between those players who are tragic in nature and those who find sanctuary in comedy, if Apollo is not the sun of our present day. How would you like to live like that? We don't really want another sun. There comes a time that we will all recognize that our sun is here to stay, because as of now it would seem as though we have a new sun every day, one that rises when it is born and sets before it dies, a comet, many comets that pass us by, one by one and once a day. The play between a tragedy and a comedy is our attempt, -for we are all good players,- to find what is beautiful about the night sky and the day sky. With four moons to circle our earth we will no longer be weak when we attempt to unravel what is fair, and we won't need to question what is fair at all, and perhaps lemon juice is sweet! With four moons we will find no need to question what is good, constantly ending up with more questions, more fear, because four moons will harbor the weight for us, like the paperweights placed on every corner of a map,- as the map lays flat on our desk,- paperweights used so the map will stay still, on our desk, upon our assessment. *To be recognized by the universe*: something our earth could not accomplish with only one sun and one moon. Something our earth could not accomplish with two or three moons. It would take four moons for our earth to be recognized by the universe because we have four directions on a map. We have not yet been able to find enough insight through the assessment of this map. The evidence is in the fact that we still haven't discovered what is fair because the boundaries of countries keep changing on a map of the earth. So let's unfold this map. *So why not make a map of the world?* I can envision a earth with a square map in the center of it, aligned with the equator of the earth, where each corner reaches out to each of the four moons that Fourier said would come to circle our earth. We've established a generic map of all of the continents on the earth, and all oceans. We adapt it while we go through the ages, because territories change, land claims, name claims. The earth with all it's landscape, terrestrial landscape, has never become a part of the map of the universe like the sun has not. The earth is no foreign planet and it does not have to be. The open ended map of the earth where the continents are visible is the closest we have gotten to a map of the earth as it relates to the universe, it is called a world map, and the space in the middle of the pacific ocean, on that map, always remains unaccounted for and yet there is no other planet, no universe on that map. On that map we only make one accommodation and that is for the size of the earth and the shape of the earth.

      There is a diagram to size of our solar system and in it is the sun. Because the sun is too big in that diagram it is not even incorporated in the diagram fully according to it's logical size compared to the other planets and our moon, which do fit. The earth and the moon make it into a chart of the galaxy, and of the universe, together, but the sun does not make it into this chart, I have never seen it fit. This all may mean that our current sun is no competitor for our moon, speaking mythologically, because the moon in that diagram is to size, as compared to itself and the other planets in that diagram. It seems as though the earth will not make it into a map of the universe with the help of one sun. The distance to the moon from the earth can be measured in kilometers, but not the distance to the sun from the earth. This is why the moon can make it into a map of the universe with the earth, and the moon has a mythological origin because of a reality like that, a mythological origin that the moon, Dionysus, can rely on, to win praise from the people and acceptance, to be a accompaniment to the night sky that no one will refute, which some people may wish to celebrate, and what a little black cat named Rak says: *be enchanted by*. When the concept of "to make a map of the earth" arises, and it Has now in the modern day, this may simply mean that we need to make a new map of the earth very soon with new boundaries that we have found to fit. The maker of this map of the earth will have to remember what a map from antiquity looks like, and the reason behind the need for a belief in a heaven and hell, and the need for a afterlife for a woman. So will that map maker remember a woman? We have mapped the entire universe, a never-ending distance of unnamable stars. But knowing this would make our universe seem very small to us: can our celestial sphere seem big enough to compete with the universe to us? That would mean that we are obliged to a old farmer's almanac, the same celestial sphere that the moon is also center to, the moon our ally in making a new map of the earth. Perhaps our universe seems small to us because we look at distances between celestial bodies as light years. If we decided to remain using kilometers of distance as measurement in outer space we would have a map of the universe and we would be able to travel distances in the universe in a tugboat, which means that in outer space travel we would not go that far away from the earth and the moon. Let us extrapolate on a map of the earth. Let us benefit, and be bountiful, and then maybe the idea that a star is a species, and does twinkle, can benefit us, then it would not be so scary to see a animated night sky. Four moons should be our paperweights when we are looking at a map of the earth that we trust. Moons are not light years away from the earth and our moon is not light years away from us. They are kilometers away from the earth. Otherwise how would the young girl talk to the moon? Four moons of ours will each be kilometers away. This alone will make our earth a reasonable distance to be mapped on to a map of the universe.

No woman could have seasickness whilst knowing her planet is a part of the map of the universe. Because we were not afraid to lose a sun if we had to. But if we were more concerned with playing pool with the planets, rather than just deciding to make a new map of the earth, a woman would get tired, and a woman needs a afterlife. We were not afraid to make a map of the earth from the viewpoint of looking up,- whilst knowing for as hard as we try we can't possibly go up on a map of the world because on a map of the world northbound eventually become southbound- and that map is the map of the celestial sphere.

If we can't all agree today and decide that Apollo is our current sun and will be and should be forever, meaning that we have not yet decided then that Greek tradition will never die, if we can't decide today that our sun is actually Apollo himself, then we will have a moon wedged at each corner of our earth. Not only will we have a dead god,- the face of the moon which we see every night,- that god strung on the cross, who always begins his transit around the earth from deep east of the earth at the 3 o'clock on the face of a clock, and he is housed there, that is his house, but we will also have a old god as Apollo. He will settle anywhere around the earth as he leaves us to become less valuable as our second moon, but remember, the moon never came to circle the earth this way. *And no one should steal the seat of the moon.* If our sun could not stay with us, to stay with us every day, it would be because everyone wants to see a female sun win.

Venus, our planet, our star, our Greek god. There has been no fresco made about her, and only has she been made idol of in ancient traditions, but not in Greek tradition. Throughout art history we have painted her as if she is just a woman, even though she could be the eternal mother, and perhaps someone or history was not wrong in not worshipping her because in space she may have a unique orbit around the sun. Perhaps she is the perfect pair for Apollo because of her unique orbit around him, but in this modern world, because of the industriousness of someone else, every man and every woman who is a feminist may want to see her every day as the center of our daylight filled days. Venus is the sun of a desert filled earth and every man and woman in the world may want to argue with her but it is better that she and her orbit and her sun-shining remains peculiar and she remains as a powerful symbol in our night sky. Then she will remain taboo, veiled, and a secret, and a mystery. Then she can be a planet and a star. And if Venus was our sun, and we could admire her then, then we would be arguing with our moons because Venus will be a sun who could not keep up with heliotropism: Venus does not have a strong enough gravitational pull. If Venus was our sun of every day we would be arguing with a totally different celestial body, a moon, and a star because Venus would never lose a argument. She would welcome propaganda, she would welcome the news, and still she would never lose. Apollo is lonely enough to be our solumn sun. In a rational world Venus could be a sun. In a irrational world Venus could be a sun. But Venus is a art historical figure. And there are sculptures to be found, yet, of Venus, and sculptures to be made yet of Venus. It is better that Venus is a sun of a future we will never see, and then the idols of her that we do find, from the past, will remain valuable.

Dionysus will always remain the youngest moon in our night sky. Whether or not he died first, he died first in our night sky: he is our dead god. Apollo will be our old god, a metaphor for a sun that we believed in once before the age of modern art. The god of the underworld, though we never saw him shine as a star, he is Pluto and will travel to earth becoming our third moon, and the god of war, Mars, will come to circle us last, after his last argument with the female sun. The order of which every one of these four stars will come to circle us is based on tragedy alone because even after everyone else will readily understand the loss of a sun, Dionysus, our dead god will never understand the loss of a sun. We must speculate that Mars will argue Venus as he comes to circle the earth, and Venus will take the place of our sun if our sun decides to follow the passage the moon takes around our earth because Venus is the eastern star, the only star that rises and sets, something that will cause mars to circle Venus like a binary star to recreate a old stupid war, nothing more than the old cliched idea that men are from Mars and women are from Venus, and as enticing as it may sound to see a war like that during the day, we need a war like that to stay alive in our night sky, because it is a mythologized war. Mars will argue Venus and then will lose, immediately, to become our last moon because in the advent of modern-day technology woman is no longer crippled, even the feminists are men. No one could question the idea of a female sun, especially a star named after a god who is supposed to be a representation of beauty. The one moon that we have will be center to a chain of events, and that will be the kilometers of curved distance that each moon solemnly follows around our earth after the lead of the moon. Dionysus will become someone we will not make idol of, but someone who we can worship: alive. Dionysus to us today is dead but He is the god we want to worship, not Apollo. He is almost not dissimilar from the idols of Christ we carry in tradition, he too is the god on the cross because he defied the kings of Greece. If Nietzsche said that we may one day have gone so far past our belief in god that we can worship rock, like a rock, we could worship the moon. Though Nietzsche meant we as good Christians could lose ourselves to worshipping a diety made of stone, if we did worship the moon as Dionysus we would be doing no different, because the moon has won this mythological war as compared to the sun, because if we compare the origin story of the moon as derived from mythology, to the story of the sun god Apollo, the stories about the sun god Apollo are few, his line of descendance is vague: we are still trying to figure out who the sun is while we worship a moon we are already comfortable with, with the name as, the solumn moon. When the moon is Dionysus to us as if he could do something for us then, those times are somewhere we don't want to be for long because the moon, who we do admire, he is dead, he is like a memory, a ancestor to us, and Dionysus disgusts us because his roots are the Greek tradition. This is why Dionysian celebrations are disgusting, as extravagant as they are, with costumes and adornments and finery, sometimes we do not know it ourselves if we are living the tragedy for as happy as we are it is like a tragic event happened to us at some time in the past and we've never been able to be rid of it, because for as much as tragedy and comedy are alive, this sentimental farce, which is that the sun is dying, cannot compete with our love for the land: Canadiana, because in our tradition the sun is not dying at all. Nature rattles and shakes so that we could remember a god that made nature god at a prehistoric, pre-biblical time but because Dionysus is a pre-biblical identity of a god, we have to be careful how we rattle and shake in the forests of Canada lest we not be accepted by the land and be rejected as we become those who don't believe in god.

<u>Chapter 2</u>

Hell

Dionysus is a young god and he lived among men. How it came to be that Dionysus would lead the darker side of what is a tragedy and what is a comedy, has to do with the fact that his personality outshone the other gods in the eyes of the heretics. He won himself a place amongst the gods on Mount Olympus because he won the heart of the people who were in the valley. It was not known then that he would make a pantheism a monotheistic religion. And he was just a man. Before his arrival there was only tragedy, there was no lawlessness or nuance for anyone because everyone who did break a rule was always sent to Hades. In Heironymous Bosch's triptych of heaven and hell and the earth, called The Garden of Earthly Delights, painted in 1490-1510, Hell is filled with those who broke the law and were sentenced to make, what it seems to be, musical instruments. What sin was it that they committed? Was it any of them? Was it all of them at once? Was it of the likes of the religious mood as Nietzsche would say, with which they spoke against divine law? Not everyone makes it to hell, so to make music for the gods the sin committed must be for the faint of heart, must be a sin against divine law, must be a sin against a benefactor, must be a sin against art itself. If one acts against divine law like they are erecting a effigy of a god, -what is most probably the greatest sin against a medium of art, marble or stone or clay,- what is a better way, for a person like that, to have to appease the gods than by making music. And what better way is there to have to appease the gods, than making good music from unexpected materials so that the poor souls could appear as jesters in front of the court and be made a joke of by divine law. The heretics never stopped worshipping a man strung on the cross. They never stopped worshipping Christ. They just simply stopped worshipping his image. Christ on the cross was a effigy, and if hell is on earth, we've fallen prey, become seduced by the sound of music so much so that it is better and worth for us to not worship a effigy so that we can create inventions. And when we do take part in Dionysian celebrations, in fine dresses and pantsuits, we wear masks. Some of the masks that we wear are frightening, hideous, sad, dreary, smiling. Who's to say that we are not celebrating the face of a god we worship, of who we don't have the luxury, out of all the riches we have in the world, to remember what he looks like. When Dionysus arrived mysticism and craze from the seduction and desire of music caused him to become the learned man, preoccupied with questioning the law, preoccupied with questioning law all as a response to the sound of music. It was simply fate. Apollo was the god at the time who governed music. It was said he played a musical instrument better than anyone could, and it was Apollo himself who was the law, the sun god, reason. But Dionysus used ecstasy instead of reason to conjugate music, developing a gathering of people who participated with him in enjoyment that was heretic but that was nowhere near to being a act against divine law. Following Dionysus was heretical because he created what was never before seen, a actual cult, and a cult without a leader. He created a cult with no worship: no idol worship, no monuments, just the sheer enjoyment of music and the sovereign landscape, so there was no broken rule. Pantheism became a subtle enjoyment of the nature of the planet: mud, rocks, earth, plants, animals, clouds, rain. No longer could the Greek Gods be the pantheistic religion they represented high up on Mount Olympus because people found enjoyment in worshipping nature, in estimating that every rock and tree is alive and equal, and so the worship of nature became pantheism and no longer was there a pantheon, which is not a good thing. And as the people in the cult of Dionysus learned that every living thing and non-living thing according to ocular evidence were all equal, Dionysus passed forward a law of equality on earth amongst the species, for that is pantheism, and that law was reasonable, and that is what made Dionysus better capable of decreeing law, the only one law that he did provide and was followed and yet this was not heretic because it was the first law ever created. And this law was supported by the sound of hell, the music that came above from under the earth, from hell, where the poor souls made music to appease the gods out of instruments that could be made easily, instruments that could create sounds like the sound the clacking together of two stones makes. But it is still unclear as to why making music, one art form, -out of all the other arts- was a service of punishment for defying the law. Why was it that the creation of music appeased divine law even though the artists who pledged their souls to one of the other art forms could appease divine law? As artists we are all subjected to the whims of a benefactor, but that does not put us in hell. There are no more artistic mediums to make, except of course if you are a musician. The question to be asked is, through the use of which art form can we make a man appear most foolish, most insincere and most unworthy of forgiveness,- because in ancient Greek tradition you can get out of hell if you can be forgiven over the time of your stay-, so which art form makes of man most unwilling to repent. Our silly attachment to newer technologies, technologies that we think could lead us to new musical inventions! This new and ever new music has provided us with a accomplice who is the new instrument, something that the divine court can only laugh at. The artist has no new medium, the dancer has no new medium, the actor has no new medium, the writer has no new medium, but the musician always will. The artist has a entire database of art to support his pleas out of hell, which is that since the death of modern art the artist labored over the recreation of the myths of our gods in search for reason. The dancer's plea is that he, up until modernity could not have looked foolish whilst dancing. The actor can't look foolish when acting, he can only act funny and not vain, lest the court give birth to another farce in the womb of hell. The writer, the true hermit, mostly goes unnoticed, and besides it is not heretical to write about worshipping a god who has a face that cannot be remembered, and it is not heretical to write about politics because propaganda was never a crime. But the musician, the musician is apt to create the mood, and yes, the religious mood, someone has to create the religious mood. The actor never needed a prop to perform in front of the divine court, and that was sincere. The dancer needed music to perform in front of the divine court, and that was sincere. The writer needed lead, and that was sincere. The artist could make a ready-made out of found materials and so he was the closest, when he did do that, to become the fool, and to become the fool in front of divine law. The writer could never write without a tool to write with, and so he could never look foolish in front of the divine court.

The fine artist couldn't work without at least one medium, and he needed one, and the musician couldn't live without his instrument. The dancer couldn't live without his body, not the actor, nor the writer, nor the artist, but the musician could live without his body so long as he had music, much like Orpheus who in one famous modern painting called Orpheus by Gustave Moreau, painted in 1865, is depicted with only his head attached to a kithara, and he was a infamous musician from the ancient Greek tradition, the most infamous. This one image about him may very well describe one of his passages to hell. The musician is the only artist who can make a sale, that exchange means he will continue to make beauty for the world, and live forever, while being laughed at by divine law at the divine court if he breaks a law and ends up going to hell. But the musician was the one who broke the silence. Because where there can be sound there can be silence, and silence is a law. Next will be a similar fate for the artist, for he became a jester when he started to make ready-mades, -let's try to prevent that. The artist came close with collage, to making a ready-made, but it is the ready-mades that sealed his fate that he too wanted to question reason so certainly that he too will be willing to give up his body to make art, but that art will be music. Artists of all the fields will all decide their fate this way, so that every artist of every field will be laughed at by divine law, when they all choose to make music. And that will begin the age when the god of war and the god of aesthetics will change tragedy forever, if we have not yet accepted that the sun is our sun to stay. We will learn the simple lesson of the art of nuance from two gods who are not lovers, that's preposterous! Mars will argue Venus. Now, a sun that is a representative of the female sex is a no-no if one is a celestial body that shines during the day. Venus will always be a female in our night sky despite the fact that most celestial bodies are male. Venus presents a string of stories for us that make the night sky timid and ferocious. But for a star that cannot become a planet,- which Venus would rather not have because she is the oldest idol of a mother, even though her take on the world is of a aesthetic- a man's world has won, because to the world she is a planet. Her take on the world is a demand for the appreciation of what is beautiful over what is just, what is fair over what is right. But she has been made idol of! She is the representation of a beautiful woman throughout art history but nowhere do we find her being admired for her ability to give birth except for what we see in a well-known sculpture. The oldest known sculpture of a Venus which was found somewhere in a cave or trap was of a enormously large breasted, heavy bodied, misshapen woman. But we can't believe that sculpture is authentic because the cave from where it came from is not, this Venus cannot compete with Uro the bull but a Venus can. Every woman in the paintings of the renaissance in most cases was a representation of flesh, Venus was no more or less, but she is the real eternal mother, not Hera, not Gaia. Gaia was a representation of a woman who could phantom the earth, be flora, be fauna, she became the representation of the eternal mother as a earthly womb once, during the time of Hieronymous Bosch and his Seven Deadly Sins and the Four Last Things, painted in 1500. *Gaia would have been known, today, as the eternal mother, if Hieronymous Bosch did not believe in the eastern star.*

     Luna is the sun that became a moon in our earth before layers of density, of soil and rock and crust built a foundation of external landscape of the earth. Luna, that is to say if Jesus is a girl, is the inescapable reality that the earth is not a mother, but then everyone asks just who the earth is. Why is it that we need to know who the earth is? What gender the earth is? That is one of the oldest idealized investigations. In the mind of a small group of wanderers the earth was a woman, once the idea arose that the earth could be cultivated. The moment they realized the earth could be mined, the earth became a woman. The earth gives to us, or so we think. But we really take from the earth in exchange for existence itself. The earth does not give us anything. But it is not sentimentality that we are trying to battle, for the earth certainly does speak, produces sound. Sounds that come from deep within the earth crawl up to the atmosphere when we hear them. But in no way should the earth be equated to a woman, because Luna is a girl, and a moon now, and she is in no way the earth. It is because we do not think the earth has a center that is solid molten rock that the center of the earth can be a moon. The reservoirs we've investigated within the deepest pools of water, within oceans, are in no way as deep as we have imagined. Like Hieronymous Bosh's painting of The Seven Deadly Sins and the Four Last Things which was a rendering of one cycle of our earthly perpetual calendar, in that painting the center of the earth does not seem so far away, and just because Bosch was not successful at creating a biological classification system of all life in that painting, simply because the Seven Deadly Sins and the Four Last Things is not a happy enough painting, does not mean that in this painting one year has not passed, and that he was unsuccessful at making a map of the earth: we are not wrong to think when Fourier speaks of four moons and one earth he's referencing that painting, because there are four corners to that painting, and in each corner a story is told, as if counter clockwise, from corner to corner, in that painting, we are seeing a story about the lives of some very rotten people in two days. We have a center to the earth celestial cycle,- which surrounds us as it turns around us,- and that center is a moon. That moon is no earth and it is not above the earth's curved surface. It is the only celestial body that does not resist the gravitational pull. When I look at Bosch's painting, the Seven Deadly Sins and the Four Last Things, I see the first chart ever made which outlines the relationship between the phantom of the earth that is our flora and fauna, governed by Gaia, the moon inside the earth, Luna, and the space between the two which is Pluto's domain. This painting is the first chart of the earth ever created which implied a extrapolation to the earth, the beginning of making a map of the earth, as if Jesus himself, sitting in the center of the earth, can see past our earthly atmosphere, and view those four moons,- the four stories at each corner of the painting-, that will come to circle our earth, because of somebody else's industriousness. Venus orbits the sun. She sets with the sun and rises with the sun. She is the eastern star and the morning star. During the age when Venus becomes our sun and Mars our last moon, with no reference to concepts the psychics use like planetary occurrences which the psychics believe cause grief or joy on earth, Mars will battle Venus for a reason that Dionysus didn't ever battle a sun because upon achieving pantheism, the only religion that could be made by man during a time that he lived through, he died. Venus and Mars will argue one of the profoundest laws we have, which is the law of language, something that could not be man-made like music is made by the hands of men clacking stone. This law is what concerns us most when we are communicating with the ones we love. When we make a mistake with our words when communicating with the one we love, something that can cause us to break up with the one we love, we finally recognize the tender law of words. Now, Mars will argue Venus, like Dionysus argued Apollo, and Apollo argued Hades and Hades argued Mars and so Mars will argue with Venus. All these arguments were a result of one certain tragic event, which is the death of a star, but Venus could remain our sun for as long as we want or need, or for as long as we should have or could have a sun, because after the sun god Apollo has gone, there is no other sun we can have. Venus will always decide to agree with herself and mars will create propaganda and this is why Venus could not cause us to fall prey to heliotropism as she begins to circle the earth clockwise. And then argument, the root of all comedy will unfurl right before our eyes, and it won't be funny, because the art of nuance can only be learned from the example of two lovers. Our dead god was the king of comedy and the tragic players have a keen desire to win. Those who will follow Dionysus will always follow Dionysus but will not argue Venus because both those gods are aesthetics. Mars and Venus will fight to keep the tragedy alive, which by then will be the idea that long ago one solumn moon could count away the hours and days and years to which the sun would die. From Venus, a woman who is not his lover, Mars will learn how to treat a lover, which is very badly because throughout art history we have treated Venus, the Ancient Greek god, very badly. And it is the god of war who can ensure that everyone will obey that law, the law to treat a woman very badly, the first time in history a law made by a woman is decreed by a man, which is to treat a woman very badly. It would seem then that Mars and Venus agree because they both want to ensure that that law is implemented, and Venus will agree with that law because she is a idol but no one has yet equated her misshapen body, her ability to give birth, with her beauty. The two gods cannot agree on one specific thing, and they never will, and they will not agree on the basis of the art of nuance, which is that though our relationships are built on love, and only are they, if you are in a love relationship, the argument can become a war.

Dionysus was humble. As Mars comes to orbit us and becomes our last moon he will be the last one to fight a sun but we already know that he is not humble. We will not know when the argument will end between Mars and Venus, and if it will end ever, which means that Venus can remain a sun forever and that is when contingency and all the word entails will be thoroughly experienced for all, because there is no mitsein between Mars and Venus, *like there could be between Apollo and Venus.* When one is in a relationship where there is a lot of argument the relationship can last forever, but in argument, if the two people involved are not meant to be pairs of the opposite sex. If the female sun does not die, we will forever have a particular pattern that proceeds much like the shadow on the moon does, forever on Mars. But that can never become a farce that does not exist alongside something we almost don't want to call a tragedy because we have never seen a contingent quality in a tragedy, and the shadows that cross along the face of Mars, in this future, will not be counting away the death of a sun and a tragedy is not meant to be the winning play, the play that wins everybody's heart who has seen it.

The constellations may not be species but when we are describing them they are a simple grouping of species: the constellations are a grouping of stars. We can compare the grouping of stars to the groups of musical instruments. Most stars in our celestial sphere make up a constellation, as if the music has come together to produce a orchestra, a string section, drums and brass and piano, but Venus is part of no star constellation, she is one of the lone stars in our night sky and she is magnetic and successful the way she is. The waves that crash against the rock by the sea shores all around the world also create a pattern not different from the coming together of groups of musical instruments in a orchestra, but we have only one tide, and we should only have one moon so that the tides do not change and Venus does not fall into a simple grouping of species: it is a secret that Venus is a binary star with our sun, and the night sky keeps that orbit, that taboo veiled. Mars will learn from Venus just how to enter into the collision of the sexes, -in a hypothetical future-, a lesson to be learned from him and his very own wife, but at this time, at this one time in history it will be Venus and Mars both playing the role of man and woman, and that should never happen because in between the two is no mitsein. As for what is contingent about this argument: we know we do not need sunlight if we have the light of four glowing moons, because it is not the sun that brought light to the universe, it was stars great distances away. How life on earth is supplemented with the energy to grow is a constant and continuous debate between what is magical about our world and what is realistic. For so long we have lived with a sun. Many have enjoyed warmth and light from the sun, but does the concept of being able to live in a afterlife, - where there is no sunlight, where there is dampness, and what we fear is death-, compete with the day? Only because there is no passage of daylight into the corridors of a world that is underneath us? Only because there is no sunlight throughout the days that do pass in such a world? And is that why we think it a place we would never agree could exist, simply because it is different there? It is not hell, or is it? Is it the afterlife? Perhaps there Is light there and through stained glass windows shines moonlight, and perhaps this place is not even underground, there is earth and trees, forests and rivers there, but there is no sun. Perhaps it is a castle. Perhaps what everybody thinks is the dreary and damp hallways of a place where no one knows where it is, but everyone is afraid of it, maybe it's one of the castles we have around the world. Maybe it is a castle we have in Canada. Why is a place like that not admirable? A place where the plants can grow without sunlight and flowers can bloom. We may not live with a sun forever which can mark off half of a twenty-four hour clock, but if we do indeed embark into the future with complacency and disregard towards needing a sun, and yet we do have one, it would be like the moon encounters shadows with a solumn complacency, knowing that one day there will be the death of the sun in a farce only the moon can laugh at, Melies' moon. With a sun like Venus, no one would want to live in a castle.

If we were to make a new chart of life it would be inevitable that we would have to incorporate two different types of stars into that chart, a star that shines during the night and the one and only star who shines during the day: we have two stars in the sky, one is the sun and the other star is named a star. One star shines during the day and the other during the night. However, because the sun and Venus are binary stars, *as if the mythological, true, duality is between a star that can shine through the night and a star that can shine during the day,* instead of the sun and the moon, because we know the moon is male, perhaps it should be that we choose Venus to represent every star in our night sky, on our chart, if we simply need one example of a star that shines during the night, on our chart, to name the species. Because of the inability of us to avoid incorporating the sun into our chart of life as a star that shines during the night, we created a binary system, in this chart of life, a fork in a tree which was just a simple grouping of species: *stars.* It is undeniable that one star in our world is very unique, and that is the sun, but we will come to find that many stars are unique, who have their own names and mythological identities. In a stupid world I decided that only two stars should be in a chart of life, and if one star is chosen to represent all the stars that shine during the night and if that star is female, then even the stars have been pinned down by gender in a chart of life. But a star cannot attain the top rung on a ladder of life because it exists too far away. In the era of modern art, where I assumed that our star Venus is a mate to the sun... How can that be so? When during the era of modern art I saw a beautiful metal object that you can hang off of the bricks of your house, which was painted in blue and yellow and was in the shape of the sun and the moon kissing each other goodbye, as the sun was about to rise, and the moon was about to sleep. In the sculpture the sun was male and the moon was female and they were mates.

We come to find that any star that has been noticed enough by us, -for us to want to think about it at all-, will have a name: as for this species, the star, for us to know if it is on the top of a ladder of life is another question, but what is a more pressing reality is that if we are to include one star into our chart of life, that star will have a name too, like the north star, like Venus the eastern star and the morning star, so why would we pick Venus as that star?

Our moon was once a binary star to the sun but because currently our moon is named so, as the solumn moon, another interesting dilemma has arisen, which is that could our current day or specifically could our oldest recorded object or artifice which captures the image of the moon, - or a phrase about a moon- be the first day when the earth had a moon and then was it that day when the ladder of life gained a order, and then so man and woman do not head this biological classification system, and the sun and the moon do. But dilemmas like this exist because we keep believing there is a older artefact which will prove to us that we've existed longer than we know to have existed so far. And we believe the story which is about how old we are through the investigation of what is a species and by categorizing species. Another interesting question is whether our current moon which was once a star that twirled with our sun created the fork which is category, because it Was because the moon did decide to orbit our earth, by will, that the first night and the first day was created, as if the weaker of two binary stars will always do that in a chart of life, because every new category created in a chart of life is created in response to argument. Venus, a star in our night sky who continues to compel us to see it as the only star in the world, and in the night sky, is younger than the sun and has not yet reached full magnitude, the star ages as it orbits the sun. For a period that will seem like eternity we will have two suns in our sky, during the day, rising and setting together, like Venus and the sun do now, but Venus now is only a tiny twinkling star. But the time when there will be two suns will only feel like a lifetime, a eternity, even though it will only really be for one sunrise and one sunset. The following day we will have a new sun in replace of Apollo, and that sun will be Venus and this will change tragedy and comedy forever. But who will rule the day, Venus? Or Aphrodite, the goddess of love and beauty? But we know that the goddess of love and beauty will change tragedy and she is the perfect woman to do it because out of all of our Greek goddesses she is the least vain. Mars would have called her Aphrodite, through a worldwide pursuit to defame her through the use of propaganda. "Venus is a better name for a planet", mars would decree through propaganda. "why don't you show yourself", Mars would proclaim in propaganda, making Venus to appear unfair in the eyes of the world. But as a last moon who will come to circle our world, Mars will provide a truthful competition with Venus, because Venus is our Boticelli's Venus. Venus is our old sculpture which we did not find in a cave or trap. How about this, Aphrodite never ruled a day, but Aphrodite is not vain, and she is a good character, but Aphrodite is not a star in our night sky, and it is Venus who all those paintings were about throughout art history, and that is what Mars would try to take away from a sun that is female, a phylogenetic tree, a career. A tragedy is built on the main character of the play chancing upon death due to reasons of retribution or cheer idiocy. There is never divine retribution in a tragedy but there is in a comedy. Venus will bring tragedy to the day, as Apollo did, but for the first time ever, the daylight hours will be in patronage of a woman who will change hedonism to something more than the involvement in controlled pleasures, she will make the daytime hours then, part and parcel to a motivation of another kind. Venus, the goddess of all that is beautiful will bring beauty to the day, in a much more flowering way than we had seen the day before. Before we had order and structure and beauty in it's place, then we will have beauty as law, aestheticism as a culture, during the day, and we will want to find refuge into the night after day, only so we don't have to be so beautiful. Venus's law will win. Venus's law will encourage us to admire the Dionysian parades and crave them where as Apollo, because his forte was strength and endurance, could not possibly drive a populous to desire weakness, like beauty can. But though Venus with her admiration for all that glistens and echoes of a structural order based on aestheticism, will drive the artist to enjoy the night more, and that is good, but, it will be nothing new for the artist, it is the civil servant we want to rule over, and the civil servant will desire weakness but he is already weak. Venus will be the heir to a long lasting tragedy and the competition as to who's rule and law and reason is more fair and practical and effervescent will be between Dionysus and Venus. Venus is a woman. When it was that tragedy was owned by a man, was when Apollo became our sun and Dionysus had to measure his rule. Venus will have Luna to circle her. The moon inside our earth is named after the very first Greek god, That Luna will circle Venus as a disciple, because Luna is the star that is inside our earth. But two women never fight with each other. They may have differing views but they will not war with each other, but *we should never fall prey to heliotropism*. Luna is a young girl, so she will not resist the charm of Venus. Venus will war with Dionysus for the sake of settling a long ago dispute of just what it is and was the birth of tragedy. *The purpose is to discover how we can find morality and abide by the laws that do not limit us but also do not recreate the past.* And this past that we want to avoid recreating as it always is our worry, is our individual, unique, constant, nagging display of emotions that portray a weakness in us, which is the doubt of if we are infact doing right and what is right. Do we want to know if our endeavors are right or just? And only a constant war can teach us this, but that was, is the past recreating itself. We don't want to war over what is right anymore, and only the investigation of fear can take us there. Dionysus will always win even though there is a equally powerful man to fight against, because Dionysus is the one that made the law, which is to worship one god instead of many. For the years, centuries that Venus and Dionysus war we will learn things about beauty and weakness that we have never imagined we can, but the root to the birth of a tragedy is when one man warred against the man that led by example. Venus can speak for tragedy in a different light. She is not a man, and she is superficially seen as the least virile woman of our gods, because she is so beautiful, but in many ways she is the most virile woman god, because for someone with such a keen eye, unlike Apollo, she takes part in Dionysian celebrations because she is a woman and women roam the streets stark raving mad at night, worshipping a god that sometimes we let go nameless. This woman will make a tragedy more tragic than any man could because as our moon, Dionysus, counts away the hours to her death, he is counting away the hours to the death of a woman who most believed was superficial, which is a lie. A old god is a god that fought a war once and lost forever, and a woman can never be a old god because she will not war with another woman and she will never demand a position as a god to be worshipped if a old god exists. There can only ever be two gods in our world and around our world, like a refuting comedy, and tragedy, and none of them can be female because if so, a woman would either have to become a old god at one time in her life or remain a prevalent god, and simply because woman herself is a idol, the identity of any woman is singular and irrefutable, because she can give birth, she must never stand alone as a god to the birth of tragedy or else she would have to take the blame for morality and immorality, and mortality and immortality, and she already takes the blame for mortality and immortality, both in her: *you can't blame me because I wasn't there, in the middle of the mourning, sings Venus to Dionysus, you can't blame me because I wasn't there, in the middle of the morning, sings Venus to Dionysus.* Venus is singular, in our night sky, and the only law a woman truly abides by is to create, like a man does, even if she is one shining star in the middle of the day, and even then she will desire to give like every other woman and there will be confusion in between the desire to give and to create for her certainly, just like every other woman, even if she is one shining star during the day, *you have to kill a planet to make a star,* is something I dreamt once. Venus is a star in our night sky, she is the morning star and the eastern star, she has a mythology to her place in the celestial sphere. During the night and the dark morning, then, she can be a planet and a star. Like a story that does not end badly, she will never become a moon because she has pledged allegiance to the night sky. Venus is a star in our night sky, and a planet in our night sky, forever, but if Venus was our one and only star that shines during the day she would only be a star who would have to follow a Law, so maybe it is easy to say that stars that shine during the night sky follow no law. Has anyone ever seen a shooting star glide across the sky during the day? In our prophetic future, which will be profitable, and is prospective, Venus will desire to create, but not create anew, but rather to recreate the past, through painting through stitching and sewing. She will promote a natural inclination of our species to believe that the woman's role is in the household, as a housewife, as a mother, but not as a daughter or a sister, but she will not believe in a maternal instinct, because she will be stuck, obeying the law, and she will be stuck, unwilling to change the status quo, which means we are closer to seeing her arrive to become part of our daylight filled day than ever before, *woman needs a afterlife just as much as she needs a sister, and a mother and a daughter.* It is Venus's law to recreate the past therefore she will never become a old god and we know she can never become a prevalent god because she is a idol. There has been a idyllic sculpture made about her, which until this day has not certainly made it into the books about art history, and there have been paintings made about her and in them she is a god, and in Botticelli's The Birth of Venus painted in 1484-86, she is a aphrodisiac. Apollo will become a old god because we will no longer need to worship him but we will want to, if Venus is our sun god. Apollo as a moon, that will make him a saint! And a idol! like the sculptures of saints in hollows in our churches. But if he is the one man in the scala naturea it would be better, because as of today, the one man who represents the species man in the scala naturea looks like a old god. Apollo will never be a solumn moon. If he does become a moon he will become someone we can no longer worship because we do not worship idols, and that is what a old god is, a old god. And I can fairly say now, that currently we can worship the sun. And I can certainly say now, that the old sculpture of Venus, of who we are not quite sure as to when she arrived into the world, she can remain a god we can worship, because she Is modern. We all want to worship a man who has so many myths attached to him, -like Jesus Christ-. Dionysus's greatest allure, which was being a mysterious man of tragedy will one day be stolen from him by a woman. She will have a sword to his throat and will demand him to step down, because the planet Venus is a mysterious planet, and then she will manufacture a almost incomparable gesture of tragedy. Venus will set forth the most tantalizing scape for the day and she will prepare a world to agree that a celestial body can be a woman so long as we don't worship her. And Venus, as becoming the heir to a long-lived tragedy will be playing on a chessboard against Dionysus, but instead knowing he will win, but playing along as forward moving as she can just so not one will try to worship her, *for god's sake, do not worship the sun,* and that is tragic. Apollo was worshipped while he was the sun and he created the most profound tragedy mankind has ever endured, but it happened when Dionysus finally left his side, so Dionysus had a part in it though we must never speak of that. Dionysus became a god that we could worship - once he became a moon from being a star,- because he was the first moon we had, he could never become old. A tragedy and a comedy are quite similar except when we celebrate a comedy we celebrate it at night and when we celebrate a tragedy, a tragedy which is a result of a law of any kind and the desire of the civilians to obey instead of traverse that law, that is celebrated during the day. We need a law, and not to break it. Only a woman can make a law that is not to be broken for the purpose of chasteness. This will change tragedy forever! Because the hidden mysteries of a tragic play is always a demand for chasteness. That is all that Dionysus has ever asked for from us, chasteness. That is all that Dionysus has ever asked from us, chasteness. But he will have a powerful role as a old god, as Nietzsche would say: in the sky now, our moon is preserved, our moon is a example of a life of a celestial body that never ages anymore. And in that state, he is old, old like a young god. And his colour never changes according to the sun. The only reason why Apollo would want to be a moon, is because then he will be younger than most of the years he spent as a sun, and it would be a success for him to inch away from death to become his own solumn moon. Tragedy changed once we had a moon to count away the days and weeks and months and years to the death of a star, it changed for the first time when we no longer had two suns in our skies.

For as long as we can remember the moon was in our myths existing as a opposite to the sun. Truly there was never a time in our myths that there was no moon in our skies and there was never a time recorded in our myths that we had two suns. But the moon was a star once, orbiting the sun like Venus does today, so where in our myths is the birth of a star that was to become our very first moon? The relationship between Venus and the sun, that Venus is a binary star to the sun, is a younger relationship than that of the moon and the sun. Dionysus, our oldest god, was our first moon incarnate. He was a star circling the sun, and that was the birth of a tragedy when the world for the first time ever watched a star die. The sun which we currently have was older and is still older than the moon. The reason as to why the moon reached a state of death, a place inside our celestial sphere, before the sun, is simply because it chose to orbit the sun rather than to be the star that was orbited. Dionysus is the younger and more enticing god as compared to

Apollo, but he is not unequal to Apollo because Dionysus, though he knew more about what it meant to have order and morality, he never defined it for any one of us. Dionysus is the moon and he circled the sun god Apollo in outer space, making the sun law and order as he made the sun center to his existence. His own life passed him by only because he knew that to lead by example only causes unrefined, unoriginal behavior, and how can lawfulness ever be met without rules that are attained by genuine observation, care and precision, perfection and the desire to never cut your losses at the time that you think you are losing the most? Dionysus aged much slower than the speed at which the sun is aging, but he spent most of his time away, orbiting the sun indeed! And yes, in deed, to a ever existing pull to live like people do, to not be the prodigal sun. In deed to a ever existing pull to live under the stars that only the earth has been center to until the modern era, because he circled the sun in show of obedience, -because he is a young god-, leading by example, *that no matter where I go around the world, I always come back home*. That it is not better or worse to be the law itself than to be someone who obeys the law because if you obey the laws of morality then you are fulfilling your destiny. Dionysus never looked towards Apollo in mockery of his rule because it was Dionysus himself who made Apollo the law by making the first choice in free will to orbit the sun. Dionysus knew that the time he spent away from being center to any event, which was when he went to as far as time would take him,- around the sun,- would make him learned and a eternal disciple of morality, because he made the law: not only did a god on the cross make the law in our bible, but Dionysus, a god on the cross, made the law as a incarnation of a celestial body, that whoever orbits the sun will not forever. On the last journey Dionysus took around the sun he returned back to part ways with a very long time of orbiting Apollo. He himself was not certain he would become a moon in the skies of our earth but wandered towards earth and heard the sound of motion for the first time and he decided he would orbit the earth. At that moment he became the law because his decision to orbit the earth meant he was unwilling to wander anymore, for the moon Is with us every day, orbiting our earth in full circle the length of a month. Our current moon did simply wander towards us. It did not travel on light years, it was not on a orbit. Dionysus is our one and only shooting star. It only takes a star a day to go the distance from which we can no longer see it, so to, it only takes a day for a star to go the distance to where we can see it, if that star can put a stop to it's motion.

Apollo will travel along no light year or no orbit to plant itself in our earth's orbit. The only reason as to why Apollo will come to circle earth at all is because of Orpheus.

Orpheus was Dionysus' first follower, a better musician than Apollo even though Apollo was deemed the greatest musician. Orpheus was the first god to sacrifice his body for the sake of creating the art of music and there is a painting where Orpheus is depicted with head attached to a lyre, with his body gone painted by Gustave Moreau 1865. It was Apollo's law that Orpheus broke which is what caused him to lose his body, that Orpheus created music that subdues animals, breaking the Apollonian law which was that not all things are equal in the hierarchy of species and existents. The reason as to why Dionysus did not lose his body first was because he worshipped Apollo while creating pantheism, by only leading by example. Orpheus worshipped Dionysus since the day he was born. Apollo will follow the path Dionysus takes around our earth, to become our second moon because he thinks that Orpheus will worship the old god the day Apollo becomes him, and if Apollo gained Orpheus as a worshipper then Orpheus would have chosen a old god to worship rather than a former god: this would cause a young man to worship a old god. There is only one former god and there cannot be many old gods but there can be two. In a myth born in the modern day, we can assume, Apollo wants to give Orpheus his body back in exchange for his loyalty but that would mean for Orpheus, that he would have to worship a old god, but it is only the one and only former god who gives back more than just a body to his followers who have received a divine retribution for sins that break the law of silence. This type of retribution will always allow us to break the law but we still will be punished for it. The punishment is the loss of a body for sins committed against the law of silence. Not the loss of a soul, not the loss of a head, and not the loss of a job: this person has committed no sins against a benefactor and the lowest rung of hell is not where this person will go. For a man, for a god who loves music like Orpheus does, but who enjoys playing one particular instrument always, this type of man would win in no comedy or tragedy, he was destined to go to hell, because if he was a tragic player he would break the law the moment he played better than a god who started to organize a hierarchy of life. As a comedic player he would break absolutely no law if he chose to play the same instrument forever, however he would become the leader of a cult, and that was the opposite to what Orpheus wanted because the law of pantheism was already created, and to be able to subdue animals by playing them music would disorganize a attempt to put a hierarchy to life and all species because we do not categorize species by their behavior. This is precisely why Orpheus was the first to lose his body for his love of music because he was the first one who could not decide if there was a hierarchal order to life or if it was pantheistic. Though Orpheus chose to worship Dionysus, and to lose your body is common for followers of such a mysterious man, he was Dionysus' first follower only because he wanted to worship a former god: he was the first man amongst many women who worshipped Dionysus and he found that not only could he subdue animals when he played music, but he could subdue women. He meant not to create any law but he ended up creating the binary system, again, for what was created once was created again but this time as a cultural system: he was the first god to denote femininity and masculinity because no man was seduced by his music. All enjoyed listening to Orpheus playing the Kithara but no man was seduced by his music. Women introduced to his music were uncontrollably tempted to listen. Orpheus did not think it of a feminine or masculine quality to be put under his spell of playing music, but he noticed for himself that every animal had a female and a male of the same species and that every animal whether the animal was a male or a female of the species would fall under the spell and that amongst mankind only women would. This and this alone exemplified to him that there is indeed gender amongst species of the world, and though he could not choose between if all species are equal or if they are not equal, he discovered that the species certainly have differences that exemplify the division in gender within nature and natural life, that there is a differentiation, a binary system which creates two categories in life under the title of gender, which is male and female, a fork in the tree of life and that pantheism could not explain that fork. He learned for himself that a hierarchy to life may not be as unreasonable a explanation to a order of all life but he could not imagine the lowest rung of this ladder and who would sit there and what it would mean to sit there: Orpheus did think about hell a lot. Pantheism was a means to a culture, and it is a religion but it is in no way a explanation for why life has a order. It is only a explanation for being, for those existents that are not living, who do not have souls.

The last two stars to come to orbit our earth will be Pluto and then Mars, after Apollo becomes a moon: then it will be hades, then it will be finally mars. What is peculiar about Hades and Mars is that they will have travelled on light years to orbit our earth, unlike any other orbit in existence and something only possible, all due, to the universal grand bend. If the universe was not slightly curved the departure and arrival of these two stars to our world of plain view, would be straight. The distance that they would have travelled would be straight and not curved like a orbit. The two stars would have merely gone and come back instead. So they would arrive to become a part of our night sky, -and the places they used to fill in our celestial sphere would be vacant,- in one night, because it only takes a star one day to disappear from our plain view of our night sky: as if, if that could happen to two important planets, there would be chaos in our celestial sphere, a little black cat named Rak says: *if we made a mistake about who is a planet and who is a star in our night sky, we would be avoiding incorporating our night sky into a biological classification system of life, a hierarchal order of all things would be forgetting one possible dichotomic order, because planets can be lacking pairs but who knows if stars should, and what if the comparison between stars and planets, between celestial bodies that may lack pairs and those that don't started the dichotomy of the sexes in a classification system, simply because they are the highest in the sky.* Moons should not lack pairs. For the first time ever these two stars will make a round trip out of two light years each, and the second light year for each will come from their very own choice. These two stars will break the laws of motion to come home propelled only by the desire to want to be in the night sky of the earth, no other star has that desire other than our sun and moon and it is not only because they are all four of the same magnitude of star, but because to them the earth spells out unto the universe that the rushing of water mills and turbines, the hushing and shushing of tides is a much more profitable gain to encircle than even the entire universe. And the reason why is because these stars know that to hear motion is much more rewarding than to aimlessly travel the universe looking for the answer to the laws of motion, because once you can hear it you want to be partner to it, and there is no space in the universe we have yet found that can produce the sounds of motion except for our very earth. In our drawing, The Chart of Life, which we will make from the Canadian perspective, the pleas of these two moons, -because they want to be in our night sky,- the pleas of these two moons will cause me to not refuse to paint them in our night sky, because we can often see them in our night sky, like Venus.

The moon however may already have a pair, and it may be the girl in the earth, it may be Luna. *To look at the four moons who want to circle our earth*. If they did want to. If all of this was not a political satire, this desire of these three moons may simply mean that these three moons not only want to be a part of our night sky, but they want to be center to their own celestial sphere like the moon alone should be, but we cannot do without the light of these three planets in our night sky. We do see these planets in our night sky and they should be named in our chart of life, as planets, not as stars: as if for this moment only the expectations of a plain view into the night sky is not good enough because Pluto and Mars are planets in our night sky, not distant twinkling stars. Because perhaps no star twinkles in deep red, and perhaps no star twinkles in maroon. If the drawing of a moon, -the drawing which makes the moon a part of a world where there is the ocean and all sea creatures and all land creatures, like the scala naturea,- can make all existents free from gravity, in that picture can the moon hear the ocean tides? In that picture the moon must be able to hear the hushing and the shushing of the ocean tide because the moon is no farther away from the ocean than is god. The Scala Naturea is only ancient to the onlooker, but it's not defunct or ill-equipped, rather it is filled with more importance than contemporary art today. This picture will live on because the creator of it had decided when he made it that gravity would be secondary to motion, and if gravity is secondary to motion then the moon can even hear when we talk to it. So why couldn't the moon hear the rushing of turbines in real life, when the earth makes the sounds of industry, and electricity, even if we do not put a whole round earth into our chart of life. The moon can and the moon does hear the sounds of industry. That is why it takes the moon a month to orbit our planet so it will follow through with us every day a woman has seasickness due to motion because the birth of a mother earth can only be stalled.

Four moons in our sky will end a woman's seasickness forever! Even if this all may mean that we need a moon in our night sky, in a chart of life, and we need Mars in our night sky, in a chart of life, and we need Venus in our night sky, in a chart of life, and we need Pluto in our night sky, all painted in a chart of life. We may not ever need Saturn, or Uranus, or Neptune or any other planet in our chart of life because perhaps the night sky cannot afford it. If These planets can make it into our chart of life, then surely they will become part of a simple grouping of species somewhere in that chart called planets. There is no room for them in our night sky! Was not all this to prove how beautiful the night sky is? But I do think there will be room in this chart of life for a vagrant outer space. Four moons in our night sky will end a woman's seasickness forever. The law of motion will be freed of a guilty conscience, something such a beautiful law should have never had to bear but the law of silence was so overwhelmingly unbearable, and the thrill of inventions was too strong to overpower. Four moons in our night sky may make childbirth easier for a woman who may find childbirth challenging, and there are four moons in our night sky right now; we have Venus, we have Mars, we have Pluto and we have the moon in our night sky. The moon says: *those other three moons are not like me. There used to be a time, that you may think existed, when we did not have four moons in our skies, but those other three moons, though they are planets, with a special lens you will find that they are round like moons. However, Venus will always remain a star, if you look closely Venus still looks like a star. Now, there used to be a time that women and men everywhere on earth believed that the woman's monthly period worked like clockwork according to a special relationship between the woman and the moon, but those three moons were always there, and the advent of technology could not make us believe that the moon is not more important than those other three moons. So, we should still believe that way. But I have died, and I am indefinitely living in a state of half-life, something that a woman and only a woman can live through too. The reason why women like the moon is about rusting. When women bleed it is like rusting because a woman will lose more iron during her menstrual cycle, every time she bleeds, more often in her lifespan, more than any man. So the moon has died, I have died, and women like that, because they may have felt that they have died many times and yet have not died. They most definitely, I am talking about women, do lose a lot of blood in their life, but they do not reach a state of half-life like I did, because they have families to support their endeavors, to reach out in the world and invent, these activities keep a woman alive, women are much more industrious than men. Busy work is good for a woman, even if it means that that busy work will make her more successful than any man, because all men want to have children, and busy work is good for a woman who will have children.* If the moon exists in a state of half-life indefinitely, *like the moons says,* that was only something I imagined when no one told me so. And as for what it meant to me. I thought that the sun, -because the sun is a star,- existed on a lifeline like every other living being, a lifeline where birth is at one end of the line and death is at the other end.

At the beginning of the lifeline of a star, -where on that line, we would mark off the birth of the star,- the star may look like a tiny twinkling cell, or the star may even have a shell, something a star may always break out of to join our night sky. In the middle of the lifeline of a star a star may shine at full magnitude or if a star is destined to become a moon it will travel during that period of time and become a moon, and will never change again. The middle of the lifeline of a star is called half-life as the middle of the lifetime of any living thing can be called existing at a half-life. The end of the lifeline of a star could be marked off by a supernova or black hole. So, I thought the sun should die one day, just like every other living thing, but I did not think that the moon should die ever, as if somehow that celestial body escaped death. The half of life is where the moon stayed, which means that heaven is real! Because perhaps people don't die. But people don't live at half-life forever: the moon may breathe, the moon may speak, never underestimate the nature of a planet, but the moon does not have skin. Scientists may say a lot of things about half-life, making the concept appear a pseudo-science, but everybody has a lifespan, every living thing, it is fundamental, so every living thing must compete with the concept of a half-life, this is how we can make phylogenetic trees about archetypes and clans. I honestly thought once, that the moon was once a sun, who travelled through space once, past the sun, and that is where the moon danced with the sun fulfilling a binary orbit around the sun, and then passed the sun by to come to circle our earth forever. The colour of the moon changed from a bright yellow to a pale white, or from a bright yellow to eggshell white or from bright yellow *to pale blue. The moon became center to his own events, underneath our celestial sphere.*

    The individual uncontrollable success of a woman in the modern world, today, exclaims that though a woman surely will do her household chores and will do them well, and will have children and will raise them well, to view the woman from a modern lens is a inevitability. Woman today inevitably has become successful in a world where success is now measured by prospects and wealth and fame: in the modern day a woman has so easily gained a individual identity, not far removed from the identity of a family, but far removed from anything recorded history could prove about the financial wealth of the common woman. The busy work that woman has done in the household for centuries has now finally afforded the woman a sense of individual freedom, -here in Canada,- where now she will be accepted as a artist, if she so chooses to be a artist. And after she has gained this fame and has to face the modern world, then she will decide if all the work she has done in the household for centuries, and raising children, has benefitted her: if the modern world can compete with the comfort of the home. But we have to remember that the success of a woman in the modern day was inevitable because up until today there has been no female animal in a chart of life. Because woman has a extravagant past which she alone can depend on for emotional security, she may be a very good artist.

    To have a moon inside of you means you do not have a earth inside of you. If we had a archetype of the eternal mother, which the psycho-analysts insist we do have, but we don't yet, that archetype is surely something every woman will benefit from in the future, once she arrives in the world and is not weighed down by the dates of art history. Surely Venus is the eternal mother but one woman will have to step down for this woman to reclaim her position as the eternal mother, in the near future. And all of this is because Venus has been a muse for us for so long, but Venus is supposed to be modern: we have never seen a Venus in our age. Without one woman stepping down, and this woman standing up to reclaim her prize, then a Venus will not be a eternal mother, but one woman will only have to step down once for Venus to live forever. That woman is Athena. If we created the archetype of the eternal mother right now she would have to have a moon inside of her and Athena is still too young for that. *This is a modern day myth*: A little black cat named Rak says. Athena and Venus are two young Greek gods who are of similar age and the only competition that exists between them is a ancient birthing cycle. A wildcat named tigerclaws says: *venus and Athena are too similar gods and Venus should see the limelight first.*

As for the archetypes of every other woman. We don't really know if they need to have moons inside of them. To have a moon inside of you means you are capable of not only housing this moon in your body, but bringing a cell to fruition, almost completely on your own: pregnancy baby! That a woman can strive to attain full control over that cell, unlike any other cell in the body, *the primary oocyte,* when the belief is popular that through mantras and meditation a woman can achieve control of organs in her body? That will only lead a Western woman astray because the only cell a woman and young girl really needs to control and wants to control is that one cell, the primary oocyte. And besides, these philosophies often teach us that the body of a woman does not work like a machine, even though it does. That you can strive to attain full control over that cell, -the primary oocyte,- means the body of a woman can take part in moments in the life of this woman. The very cell that the woman wishes to control can grow and have space in it, and that cell can be shallow, and can have a thinner cellular membrane on one side of it, but that shallowness does not make the woman meek, or weak, or too vain. Because surely success is the end result for that woman. To have a moon inside of you is a gift, but only if you embrace seasickness for as long as the tides demand! If the woman feels seasickness at all may be because she is political and the topic of discussion on the dinner table should never be about religion or politics. But I never imagined before that four moons would have anything to do with a seasickness that a woman may feel when she is pregnant. And if she has a primary oocyte in her ovary for so long, could she feel a seasickness because of that? And could she have seasickness because inside of her body is a secondary oocyte awaiting fertilization? The primary oocyte and the secondary oocyte may not ever be shed from the body of a woman during her period or during any other time in her life, but these two cells, no matter how shallow they are, it could be a shallow misconception that they are too heavy. As if these two cells are strung there by a dream catcher.

If these two cells are both there at the same time, it is like they complete a convergence of moons. And if we can see this in a ultrasound, this event, of the alignment of these two cells according to each other, -as one is tucked away in the ovary and one is in the oviduct,- I would propose that this is universal amongst women, that the alignment of these two cells, under a ultrasound, will look similar amongst all women: the pattern of a sibling for the child of that mother. This pattern reminds me of the big dipper, something that should not be too hard to carry for a woman, but not for too long. And so if four moons have anything to do with the seasickness that a woman may feel when she is pregnant, if a woman should be allowed to tell you that four moons are managing some type of worry for her during this special time. If she is political, then she should be able to say that! And if a woman may believe in a ancient cult of woman, she should be able to proclaim her faith in the planets, though then, we know that she is obliged to the help of the moon. During her pregnancy, in the modern day, she should also be able to voice her opinion about Venus, and about Mars, and about Pluto, because during pregnancy, when a child is growing in the womb, the water that continues to compel the space in the embryonic sac to become larger, to sustain the child, surely could cause seasickness in the body of a woman which is as physical as being on a ship on the ocean.

Though the law of motion is hard to live by for a woman, it is important to live by for both man and woman. The ancient Egyptian gods always had it right by naming the sky a woman, Nut is her name: we prefer to not name any celestial body a woman's name, because then that celestial body puts into question just what it is that is living and is animate and what is dead, in a chart of life, solely because we've named that celestial body a woman's name. A very real situation that can arise when we name the life that is inside of a divine order, as women or female, is very specifically related to the ability of a woman or a female to give birth, to produce a child that comes out of her body. Inside a chart of life a female animal may cause us to question the gender of a flower who is inside of the chart of life with her, and we should already be sure of the gender of that flower before we paint it into the chart of life. And that is why woman is the true existentialist because the quality of her situation always brings into question the existence of other species, specifically the material reality of if what sometimes we would call inanimate objects are actually alive in the natural world. Nietzsche, the honourable man, who coined the idea of seasickness when we sail over morality, was tempting us to resist bureaucratic law. We are certain he was talking about motion when he asked if it was necessary to sacrifice god himself in exchange to worship gravity. But the worship of the earth in place of god. That exchange happened before the potent lyrics of Nietzsche, and like he said, it was unlawful. It is a question as to just which god Nietzsche was obliged to when he expounded the importance of him, perhaps the only god he ever talked to. We could have only worshipped the earth by making the earth a woman, perhaps because we are in refusal of a god that could be a man because of our confused perception of the fertility of a man, but art history has proven that in the western world we have not been confused about the fertility of a man, as we can see in Picasso's sculpture The Bulls Head 1943. We made the earth a mother because we wanted to make reparations for the years we had forgotten that a mother is a vital force. Woman is born sailing over morality, because some people in the world worship the earth as a mother. As much as we try to resist, that no planet should be female, or have a woman's name, Venus always wins, but only Venus, because she is in our night sky and she is beautiful.

The only celestial body that is light years away is a star, and the end or the beginning of the universe. If the length of a light year, the farthest light year we have, which is the end of our universe, were to be travelled by a star then we can finally end the ending or the beginning of the universe, but that star would be a child lost to a mother, and that star is young, and that would be a sacrifice that we are not willing to make. The only thing we can wish for is that the moon of other planets are just rubble, a completely different species than a star and that the new woman who will give birth to stars, will one day.

Though a moon was once a star and a sun, it is dead, no longer dying, and living only in our minds, and the moon is alive if we think it so, like every pebble rock and stone, like every gem and metal and diamond, but if the moon headed a biological classification system of life now, separate from what the sun heads, and what is a star heads, if a sun and a star do, the moon could only really head a map of the world: terrain, territories, ocean, shores, mountains, valleys, desserts, tundra. The moon would head landscapes, and all that the moon already heads. Once knowing the moon is dead it is easy to look at the moon as unanimated. The sun is only inanimate to us if we believe it so.

Chapter 3

Cooperative Labour

We can compare a lightyear to our earthly year. Our earthly year has a half-life, -a point in our year where there is no more return, when the year has been exhausted, when the most boisterous blooms have already been seen, when the sun is hot enough high up in our day that we are ready to say that the year is over, a dying point. Our earthly year has a half-life, a time when we have not missed the signal which is that one hot summer day, the sun is so large that we can almost see a god strung on the cross in the face of him: when we can see the sun reflecting the moon, that is the dying point. Something we recognize to be the ending of our year. That the year and every living thing will find reason to grow again! Our year comes to grow again after it's death. Within the length of one day our year begins and despite the fact that the leaves will change colour and fall off the trees soon, and then snow will come to blanket everything, once the year has begun to grow again, that day in a year will come to exist during a time when the soil is fertile. A light year comes to completion for most stars despite the fact that there is a time period in the lifespan of a star, which is a half-life, where a star can stay, alive, standing precariously on a ever present life, that is if we believe that the stars we have in our sky will always be there and will never change. If the stars and constellations in our celestial sphere, our entire night sky, never change, then our entire night sky is existing at half-life is what I think. If a star travelled the distance of a lightyear, if a star does travel in a vagrant outer space or in a pristine night sky that never changes -but that which within we enjoy seeing shooting stars,- and if a star must travel to fulfill its lifespan, to fulfill a lifespan where it does not need to change colour, because no one is like the moon and no star is like the moon, then there is a endpoint to the lifespan of a star, and a lightyear, but that endpoint may not be a colourful explosion, always. Someone travelled the farthest light year from where we are in the universe once, and that someone was a star.

But there is no endpoint to the type of life-span that the perpetual earthly calendar, the earthly year, exists in. Much like the menstrual cycle is existing forever, in the body of a woman, and she avoids staying forever in a weakened state after the period has arrived, throughout the passage of our year at some time we avoid death, however, have we ever seen a moment in the year which we can equate to a weakened state, in Canada? In the mid of summer we see the season change, as the nights begin to grow longer again, at a time when we wouldn't expect to believe that the season has changed at all, because then, it is still summer, and hot. In the mid of summer we see the season change. If we looked at it that way, then can we certainly now say that the harvest season is bountiful because the season takes place at the beginning of a year- October, the Limbourg Brothers 1413-1416.

If it wasn't for the nebula in Orion's womb, it would be foolish for us to be concerned at all for the universe because it wouldn't be our very own constellation, our myth of a warrior woman with great birthing hips, that gives birth to the universe. Orion for centuries has been a woman in the guise of a man. We are slowly conquering a universe that does not want to be conquered. But why did we not see stars ever so often being sent out of the nebula, Orion's nebula, the nebula in her womb, when surely at some point during history and pre-history those stars needed to find their place in our night sky. That is because the length of a common day is long enough for one star-because every star leaves the nebula one at a time,- to travel to the place where it will stay as a star in the night sky, where it will shine during the night. On the day of the birth of a star, by the time the sun sets, the star is so far away and we don't know where it has gone, and that is another sentimental way of looking at light years. A star can be born at minimum once a year, if not once every two or three years. The distance a star travelled designated it's magnitude, and a half-life that every star must have within its lifespan, plays no role here: no star is born in weary resignation to its half-life. Every star of the same magnitude travels the same distance to be born, stars have to travel to mature, they do not mature as they travel, and that is what a shooting star is. The sun sets three hundred and sixty-five days a year, the night does not fall three hundred and sixty five days a year: that is the lost day hanging over the conscience of our clock, like our constellation the crown hangs over our heads tucked way up high in the northern hemisphere, surrounded by a circle at the north of our earth, the last meridian. The first day is after dusk on summer solstice: the shortest length of night of the year, which is the beginning point of our earthly year, but because we do not recognize the night as a day unto itself stars are birthed from our favourite nebulae quite slowly. Whether or not the stars are there or not when we cannot see them during the day a star is born on the longest light filled day, this does not matter to us because we will see that star the following night. But the mother that hosts our sky cannot give birth to a child as frequently as she used to because as the sun, our sun, is dying, it is destined to become brighter and brighter. It's nearing death is causing our days to become longer, still at a unnoticeable amount for our worldly view but for a woman who is timing her births she has to wait a longer time to give birth because one revolution of the earth is becoming longer than twenty four hours. And she will give birth to a star, to be born again, to shine, in some place in our night sky, in the present, on only the condition that the day she sends out her star be shorter than the following night so that each star born from her will be sublime to the idea of darkness in the universe. Our invention, the generic calendar, and every enunciated day may change, give or take a day or two, because our sun is dying. Our stars are sublime, our constellations are sensitive and docile like a lamb, as much as they are strong and components of nothing but their own existence. The stars do not exist for us but they accompany us, like the north star does, while we live on earth. we are not small, we are not big and that is what looking up at the stars can teach us.

Orion is a woman with a belt and a club and a lion head in her hand. She has big birthing hips and a womb in her left abdomen. She is the oldest shape of a woman we have in our constellations and she acquired the wealth of our admiration for centuries by pretending to be a man and a leader in that guise. Now that we know she is a woman, the secret is revealed and perhaps the time is ripe for it because soon she will stop producing stars and another woman will take her place, and this time that woman will never have to disguise herself because we have come of age. Maternity is a longer length of time than even eternity, eternity is just one day. We have spent centuries of nights looking up at a woman who is made up of stars, who harbors a matrix within her dress. We never believed in a myth of Orion as a woman, so we couldn't embody that myth, we couldn't paint pictures of it to unravel its mysteries, but now we can. She is a new addition to the database of modern art, and the only reason she gave birth was because no one knew she did. Now another woman will have to come along because time doesn't stop but because it is not customary to disguise oneself when you are a woman today, we will never know she gives birth to stars because her nebulae will be black. The universe does not want to be conquered and nor do we want to conquer it, but we want to remain existing in order and so long as we envision the life of stars travelling away from us in light years, as if some force of nature had put Orion to it, so long as we envision the life of stars travelling away from us in light years, we are in charge of motion, so that our nights won't get brighter faster than the eventual dimming of our sun. So that our nights won't get brighter faster than our sun becomes the biggest and boldest it has ever been and before it reaches a dim finale. So that we don't blame the sun for our nights getting brighter.

    The ancient Egyptian method to putting a order to life was to equate a cooperative system of labour to all life: they believed in a afterlife and also believed that their most valuable objects would go with them to the afterlife, objects like a plow, life like apple orchards, and painted and drawn images of bulls plowing the fields. Whatever industry they did have, what little work they did do, was always going to benefit them, but in the future. There was no enemy to war with, who could divide that fertile land between religious and societal ideals, and everyone on that land believed in a afterlife, so it became easy for this ancient world to envision a ladder of life that was a exclusive viewpoint from the edge of western civilization, and though their viewpoint was exclusive to what they could see about their world while living on their land, we in the west did not forget the eastern edge of western civilization. The ancient Egyptian gods found their society irresistible. They benefitted from the fruits of their land and their people did too. There was never a surplus of goods but there was also no waste, and so in a world where there was no profitable surplus, because there was little expansion and no need to monopolize, the ancient Egyptian gods could instead believe in materializing a after to life as compared to materializing the imaginary country who was so unorganized that they could not see past the next day. They did not need to change much about their sovereign landscape in order to eat and to live and feel fulfilled, and so nature became to them, not secondary, and so they could easily compare themselves to the kingdom animalia and the kingdom plantea and all natural life. The plow which was used to till the fields was natural to them. The orchards from which they picked fruit was natural to them, and the barns where the animals were housed were natural to them, and the brick of their homes and castles was natural to them: All of the parts of a society which usually only benefit the working man to work another day were essential to their viewpoint of a divine order to life. There was never a industry or a advent of anything, -unless it was myth-, whether it was the advent of industry or the advent of cooperative labour, in their world, to arrive to separate the land. The landscape was always sovereign. Their world, which relied on a cooperative system of labour, was so unified that all natural life became a part of their stories and their religious views. And this is why they could have envisioned a tree at the top of a ladder of life, specifically a evergreen, because certainly the evergreens born on Egyptian soil were the oldest trees there and the oldest species there. Only certain societies have the luxury of being able to put a evergreen on the top of a ladder of life, societies that did not have to cut down the trees of their land to accommodate the needs of the population. Their society was the closest thing to a pantheism we can agree to admire. Because cutting down trees and clearing the land for agricultural purposes was not necessary in Egypt, we in the West can relate to this old heretical culture, because cutting down trees and clearing the land for agricultural purposes is not necessary for farming on our land in Canada either, because we have plains.

    This year, we needed to revive the old testament because of the validity of olde art history, because of the validity of the Medi-Evil era and the art that came from that time period. This year we wanted to believe in the Adam and Eve story and that Adam is the father of mankind and Eve is the mother of mankind, -we wanted to believe in Genesis,- and we wanted to believe that Galatians is about the seed of a flower. We wanted the old testament to be beautiful. We fantasized about the Ancient Egyptian gods, because ancient Egypt has a art historical period, -and Adam and Eve did too,- otherwise our topic of discussion, our single role model from the edge of western civilization, from long ago would have been Moses, and not a desert-like land, and not the gods who wore masks of animal faces on their faces, who worshipped that land once.

It is a prophecy from the modern world that the Ancient Egyptian gods equaled themselves to a cypress tree when they envisioned themselves in a chart of life. It is a prophecy from the modern world that these gods didn't cut those trees down. And it is a modern-day prophecy that evergreens all around the world stand up somewhere in some chart of life higher than the rest of life, because they are older-, not only do they stand higher on a chart of life, as compared to the flowering and fruit bearing tree, but the evergreen may stand higher than everyone on a chart of life, higher than every living thing, because he is older than every living thing.

Living within the stories that represent our stars we find that we have riddled ourselves and it is because we have imagined a celestial sphere. There are 'eighty-eight' regions to that sphere within which we have located all of our constellations. We have fit them into this sphere quite neatly and this is where we have riddled ourselves because we do not know if we did this so that we could create the zoo that is our constellations, and so have stories to live by and believe in, or so that we could have a honest and predictable night sky to accompany our four seasons and to accompany us while we live through these four seasons. Before we made a map of the constellations- from the inside view and from without- we made a map of the world, surely, and then we made a globe of the earth and the countries, and painted the constellations, animated, on top of it, as they would look from standing inside the celestial sphere. And then we made a map of the world animated with our constellations from the outside view. This animated map, when pasted onto a round globe, it was a drawing of how the constellations make mark on the terrestrial landscape of the earth, this is our zoo, however a map of animals and characters who compete to pronounce one day at a time is our zodiac. We made too many of one of these globes, and that was the globe of the view of the celestial sphere from the inside looking up. Once, and ever since the view from above of the celestial sphere was animated on a handleable globe, it was surely the first object to come out of astronomical studies that was not a work of art, and not a invention, and it was not a plein aire drawing. Clearly, it came out of someone's imagination, but nature claimed it before it could be a piece of art. This globe could have been a modern art piece only because it subverts our skies, and then there would only be one of them. If we are to make one of them now, only the accuracy of the hand of the artist will prove it to be a piece of art. Drawing the celestial sphere from the inside view is a plein aire drawing. There is no rule that a plein aire drawing must be accurate, but it is traditional that it be done outside, where the subject matter is, and landscapes, vast landscapes are a common theme in that type of painting. Our stars are part of nature, and become part of the landscape: they are the open vast space that we can draw in the tradition of a plein air drawing. In making a globe of the earth we can look at and handle and view from above, with the stars painted on it, as seen from above, as seen from someone's imagination, we make a landscape out of stars that are far away from our earthly atmosphere and from each other, those constellations are a landscape now. And as for the view of our stars as seen from the plain view, from standing on top of the earth, those stars are part of the landscape too, and the only reason as to why the creator of a globe of the stars from plain view does not need to be en plein aire to create it is because it would be too dark outside to paint it, it is the only landscape that is seen in the pitch blackness, besides from the deep crevices of the ocean.

    A plein aire painting to come out of art history which was done in the dark is Starry Night by Van Gogh. How he managed to paint that curious painting outside during the night is astonishing and a mystery. Maybe Van Gogh chose the subject matter to be the sky because it was his contribution to the map of the world,- much like many paintings from art history have contributed to our present day view of the ladder of life. In starry night Van Gogh articulated as if in speech itself that the night sky does in fact exist and does hold stars, and is a part of our world and is a part of the map of the world. It was a act of glory that Van Gogh painted the night sky, because those constellations carry with them myths, all which we have forgotten or have left behind. And in that painting too, there are cypresses, and so in that painting it is not hard to see that a season is being celebrated, the night. *And so, in the modern day, I imagined that the cypress tree lives through seasons, like other evergreens do, and that the night is a season and a event to this species.* Van Gogh in his starry night revealed the truth about the cypress tree, that this tree does not fall prey to heliotropism, that the branches of a evergreen do not reach out towards sunlight during the day, because the curly leaves and branches of the cypress tree are dancing at night in this painting. Every constellation in the night sky is a species, and every star within each constellation is too, however the cypress tree produces descendants and not every star is a descendant of a nebula: in starry night we see that the night is romantic because the subject matter of this work of art is our celestial sphere, at least every star in our celestial sphere has a family. And the rolling hillside in this painting is the domain of the cypress tree. The celestial sphere is captured spinning with stars inside it, in this painting. Van Gogh's starry night is one of the few plein air drawings of the stars and the night sky we have in our database of art history. As for the cypress trees in this painting, we do not know if they are ancestors to descendants or born in a first generation. If this painting was titled cypress trees, then many eastern European paintings from the database of art history would be competing with the landscapes of Canadiana for popularity and recognition, even more, -landscapes from around the world could have easily been forgotten if we popularized Canadian Art: Here in Canada, the amount of resources available for our people is large, perhaps a tree from the other side of the world would be forgotten in the face of Canadian industry, and in a chart of life, if the natural competition between species, including ourselves, is to not only make it into a chart of life, but to be one of the species on a higher rung on a chart of life, the cypress tree would not win the highest rung, because the pine tree would, if the competition in a chart of life is that whoever is older stands higher on this chart of life. Canada has the oldest trees in the world, that have been untouched by industry.

The ancient Egyptian gods wore masks on their faces during the day so that they could never ever be incorporated in any biological classification system as species or existents by us -*when we decided in the modern day to make a chart of life and include the ancient Egyptian gods and their art historical period called the underworld in it,*- because now we are finding dinosaur bones in Egyptian soil. Certainly the ancient Egyptian gods did want us to believe that they came before the dinosaurs, and that though we have found dinosaur bones on Egyptian soil, which we have, which according to our calibration predate any known artifact from the art historical period of the ancient Egyptians which is called the underworld, it is better to believe that the myths around dinosaurs are fables. Also these gods did not want us, surely, to put a dinosaur on a chart of life, because that would cause us to assume that the underworld did not exist forever, and that hell did not exist forever. According to history if the dinosaurs did exist and when they existed, sea level was much lower and apparently even because of their existence, sea level was much lower than the most shallowest sea, so much so that there would be no hell, and there would be no underworld, and there would be no afterlife, because the afterlife and everything that is part of that world, - for the ancient Egyptian gods,- also meant artifacts that we have found hidden under Egyptian soil. And so, much like Athena, the young Greek god who was competing with theories of Pangea, so were the ancient Egyptian gods competing with theories of Pangea! The ancient Egyptian gods wore masks on their faces during the day so that they could never ever be incorporated in any biological classification system as species or existents in the modern world. They knew that we would look back to Ancient Egypt, from the modern world, and would note that time as a time from which dinosaur bones were found on ancient Egyptian land.

    For the ancient Egyptian gods, their belief in a afterlife gave them a advantage to reclaim a chance to discover the divine order of the earth, away, and at a different time period, from a time that they knew would give to the modern world, a date, of a era, for the findings of animal bones along with findings of dinosaur bones. The ancient Egyptian gods did want us to find animal bones from their time period, but not dinosaur bones, and certainly they did not want us to believe that those animals were of some ancient descent, unlike what surely they wanted us to believe about cats. It is fair to say now that the ancient Egyptian gods and their culture which romanticized a underworld, has been recognized in the modern day, as a time that was just as modern as today.

    To avoid the day is common for those parties who have to poke and prod at plans for the future every night, but it is just as rebellious to be able to live through the day, but beware, if you are alive during the day, nature will not resist, and you have to be part of the rungs of the ladder of life. What is alluring about ancient Egypt is the cypress tree. If we are going to make a ladder of life on a circular map, rather than a square one it would be ideal to top the ladder with a cypress tree because the cypress tree has curly branches and leaves. This chart that maybe we want to make, it could be a painting of all life, and then it would be incorruptible, for aren't we in debt to paintings, still? If the ancient Egyptian gods had to choose one species to stand atop a pyramid and rule the world of the living, and the dead, and in Canada here we would choose the pine tree, without doubt we should believe that it is the cypress tree that the ancient Egyptian gods would choose to be on top of a ladder of life that they realized from their point of view where they stood on top of the land mass we call the earth, because the cypress tree is the predominant species of evergreen in Egypt, and the most predominant tree, in Egypt, and is the oldest species on that land- it is irreconcilable that some people think that the oldest species should be on the top of a ladder of life. The ancient Egyptian gods had a strong agricultural society when they did, and the only reason why they did have that was because they picked a cypress tree to be atop their pyramid of life, a pyramid of life they barely spoke about. They tended to agriculture. The cypress tree needs no tending to, so it seems obvious that the only species you don't have to care for be atop the ladder of life, because isn't that supposed to be god? Uselessness but not frivolity became the ancient Egyptian philosophy to the taxonomic ordering of species in the world. For as far as the eye can see, we notice uselessness. The only useless tree in a industry, especially where writing down information was not important, was a evergreen tree. But upon marking out the importance of the cypress tree, the ancient Egyptian gods wanted to be equaled to the tree, because though wearing masks allowed them to stand outside of all that is living, they were not dead yet, the masks that they wore during the day, and because they wore them during the day, allowed them to stand beside the tree on a chart of life, something that could have never been done before. As those who have rights to this land, rights to flowers and perfume, rights to till the soil, rights to live through the day, we could wear masks during the day too. But if we come from a origin that is prehistorical, that is biblical, after we have gotten so far in grouping our species, we discover that the father and mother of man and woman, Adam and Eve who have a myth that would be unveiled, they spent their days naked, and the Ancient Egyptian gods spent their days naked too. Because the Egyptian gods could stand on the same rung as the cypress tree, all because they wore masks of animal faces during the day, there we find and found the root of Egyptian pantheism, not much different from the Olympic gods, who sat around their round table atop Mount Olympus. The ancient Egyptian gods decided to stand beside one species of tree as it's equal, but the ancient Egyptian gods did not see any other species as equal to them because they did not know on which rung of life the other species would stand, and because everything other than the cypress tree was part of a system of cooperative labour, a wheel of life which would be understood in the afterlife, -they believed-, they chose to stand beside the evergreen the cypress, the breadwinner of their society, because it was the only species that didn't work. This allowed them to go into the afterlife when they did, and they did decide to leave their land at a opportunistic time, when they knew they could leave behind a tradition which could be called art in the future. They believed that the afterlife would introduce a divine order of life to them, where they could keep their customs, which was to wear masks of animals on their faces during the day, and then still they could be part of a chart of life, without having to reveal their faces! The Ancient Egyptian gods needed to be on the same rung on the ladder of life as one species, to gain entrance into the afterlife, otherwise they would have to rule forever, having lost this chance to wage war, because at that moment they were the strongest community in the world, they would have won any war they waged, but they decided that they would rather want to be in a play, they wanted to be within a comedy and a tragedy, because they were overabundantly rich, and they lost that war too. They did not care what species they were and are. They were not certain like we are not certain. They did not care about which species was more complex, they were pantheists in almost every way, but they were wise enough to know that death or the afterlife could be a place they could be and they were certain about if other species or all other species go there too.

  Until one twig of the evergreen tree begins to die it is living, and it never dies really, and only evergreens are this way. In one year the tree may not live through four seasons, not like a angiosperm will always. From the birth of a evergreen it can take up to ten years for it to go through it's first spring, summer and fall, because it takes a long while for that tree to take part in sexual fertilization, and it will still never experience a winter, it will only experience the night, the season night, because the tree does not hibernate. Only when a evergreen's branches begin to go bare will it start to experience its first winter and the tree will continue to experience winter from then on until the dying branch dies, but that dead branch must break off or else the entire tree will be subject to winters to come. The tree stands out in life because it bears no fruit that can be eaten, and stands out when dead because the tree, if ever becomes bare, is a result of endless winters which did not arrive to it and leave it according to the yearly cycle: the tree stands out, is unique, is ever present and that has been known to many painters who paint Canadian landscapes. The evergreen is a useless tree because this tree always stands out, in life and death, surely this species can enter into the afterlife, alone. So, it seems as though the evergreens are part of no cycle of life at all, are part of no calendar where we can remember the tree's origin or its past, but, perhaps that tree was here before we were first here. I feel certain that the stars in the sky do want to be a part of a fertility cycle that we want to be a part of. That is why I believe that the great myths did arrive to this world at the same time we did, and at the same time the stars first did arrive, but the evergreen trees and their exotic fertility cycle remain quiet towards us. Fertility is the calendar created by example by the gymnosperm, influenced by the fact that the oldest and first seed plant, a gymnosperm, has branches that never go bare unless they are permanently in winter, as if that is when the soil underneath the snow or frost or fallen leaves is the most fertile. The ancient Egyptian gods never believed in death. The afterlife was just the other side of life, and that is why a gymnosperm was on the top of the ancient Egyptian biological classification system and was what the entire agricultural system of ancient Egypt was built in mirror image of, because that seed plant produces offspring like the gods would, offspring useless to mankind, useless because they never died. But this mirror image, the one that only a gymnosperm can create, right on the land, it exists somewhere for certain on Canadian soil. The youthfulness of Canada, the youngness of Canada, is no excuse for how industrious the land is here. And here we believe in a different sort of afterlife, because we never have to leave our land to find it, the shade that is provided to us by these big evergreens is enough of a shadow to find a afterlife in. The cypress tree was the first tree that the ancient Egyptian gods noted to be a species and they decided this because the cypress tree produces offspring that are always only descendants. If this is true, from the perspective of the true north, it could and would be possible that many pine trees are older in Canada than cypress trees around the world simply because they are bigger and taller and have unfolded more branches. Gymnosperms bear seeds that can only grow to be offspring according to their fertility cycle- a much longer yearly cycle than angiosperms-, angiosperms depend on myth and myth alone to pass downwards offspring and to be the only ancestor to that offspring, like the example of the apple tree eve ate from, because the age of cooperative labour has always come afterwards, at a later date, at a younger date and *every living thing that is young depends on a true myth to stay alive and reproduce*. The gymnosperm's cycle of life and death represents the prime example of fertility because as the branches on that tree start to go bare, it still produces seeds until the last branch has 'entered winter'. Each branch of a evergreen produces seeds on it's own will, unlike the industriousness which is the flowering tree. Within the vast landscapes of Canada we can see many trees that stand still while some of their branches have gone off into the wild, meaning those branches for a short period of time if not for forever will be bare, this is when a useless tree has entered winter. Some angiosperms will die and remain not cut down by the governing forces, these angiosperms become tundra and they are never angiosperms that take part in cooperative labour in the neighborhood. These types of trees are wild.

   Angiosperms go bare in the winter in a four seasoned region only, so when their branches begin to die, they die, they do not enter a season, and as we know, all flowering plants die before winter arrives, except for the rare rose in the midst of a frost- bitten forest: a third bloom which is really the first bloom for a perfume maker. When a gymnosperm's branches begin to die, it is like they are going bare for the season because gymnosperms pass through four seasons no matter what region they grow in. What this means is the gymnosperm shows signs of seasonal change like a dying branch of a apple tree for example: if the branch grows bear and falls off, it will never come back to grow again on either of these two types of trees. On a apple tree, once it's branch has died, the branch is dead, that is one generation lost for that tree meaning what was a entire tree cannot be picked bare once that branch is dead, and in our modern day of cooperative labour we will cut that branch off.

    For a flowering tree, losing just one branch most likely means with our help it will not produce as many malformed fruits, for we prune our trees. What makes a evergreen tree's dead branch enter winter is that slowly another branch will enter winter, and that branch's leaves will fall, so it will enter fall before winter, and before it enters fall it will drop it's last seeds in spring, and live through summer, ready to be a branch that will die. And this all happens to one branch each on a evergreen tree. And when the last branch is dead the evergreen would have lived through hundreds of years of four seasons. Work is part of life. But it was not predestined that we would work in a industry other than agriculture, forever. One cannot live outside of the four seasons, ever, and winter is the last and final season of life, it is the afterlife. From a lineage of ancestors, in the life of a gymnosperm, a offspring is born until the very last branch of the gymnosperm stays intact, even though older branches have died. Who wants to store the seeds of a gymnosperm? Though ancient Egypt had the most successful system of cooperative labor in the world, the ancient Egyptian gods succumbed to the idea of uselessness, like us in Canada. That is why they wore masks during the day. So that during the day, the highly esteemed system of agriculture and industry could work by itself, and they could be dressed in attire for uselessness. To them a biological classification system was based around the reproductive ability of a species, not it's gender and not it's name. It was hierarchal because they picked only one species to notify, with the earnest implication that all other species were equal but not classifiable according to their knowledge. They hoped that all species could hide in the veil of darkness whenever they wanted to, they did not know that they could, because they themselves could not. To the ancient Egyptian gods, a chart of life was hierarchal because they picked only one species to notify, with the earnest implication that all other species were equal but not classifiable according to their knowledge, because of how equal they all are. But still a hierarchy existed in their findings because in the moment one chooses a species to be classified, and if they do not explain that the act is the beginning of merely a simple grouping of species,-

which those gods could not say because they felt that every species is equal,- noting only one species puts it on top of the tree of life. There was only one species that the ancient Egyptian gods were certain of who did not die. *And we have never seen a biological classification system that considered viable entrees like those from the kingdom animalia, or those from the kingdom plantea not viable:* says a little black cat named Rak. So perhaps ancient Egypt did know that animals do go to the afterlife.

We also have not often seen a chart of life where species are noticed by their unique capability to live on after death, and we also have not made a chart of life where we can recognize winter as a season.

Ancient Egypt could never be charted or written down in text because it was the root of what we have found today, to be a secret religion, and on earth today the nature of the reproductive cycle of a species does not claim a relationship to a ancestor because we have never put a afterlife in a chart of life. But for a evergreen tree and in comparing the tree with angiosperms we will see that the reproductive process of a tree is directly related to how and if it will have a descendent, one that may exist in a chart of life: the ancient Egyptian gods had to consider, once noting the evergreen as standing atop of the chart of life, and they were considering this before they met with the afterlife, that seed dispersal is a part of the life cycle of a tree and to bring a seed into the afterlife is a gift that is unstoppable, because there is water there, and air there, and animals there, in the afterlife, to assist in seed dispersal. And as for how our role in cooperative labor has affected how we think a species can make it into a chart of life, according to the ancient Egyptian gods, having a descendent is what will put you in a biological classification system, rather than having a ancestor.

To create a chart of life according to fertility or more specifically the harvest season and every day and moment that leads up to that season in our year, that Is to recognize that there is a afterlife and a after to life. All of the species will contest to it, that there is a subterranean place where the living can only go, where man and woman can only go and not come back from. All other species can go there and come back from there and for the gymnosperm, the afterlife is a excuse to just simply grow taller. Because the four seasons represent no ladder to climb or rung to exist on, people often think that this ever prevalent law of the perennial four seasons, -because no one has seen this law in a chart of life yet, like we have seen gravity and motion-, is a free passport to everyone: some people think they can travel all around the world and be welcome anywhere, as if we sold our seasons in exchange for free passport for others, a representation of equality of man to a species he himself chose to be on a chart of life, but when you go beyond the borders of someone else's land you still have to pay for the entrance to the afterlife.

This species was in no way different or similar to mankind. We chose a favourite species. We put a evergreen on the top of a ladder of life once because we were comparing ourselves to something that lives through four seasons many more times than we do allowing ourselves to locate our origins in our calendar and not our year,- a calendar that houses the menstrual cycle, a calendar that coordinates the birth of children from the body of a woman,- allowing ourselves at one point in time to locate ourselves within a calendar that is something different from the generic calendar, because at this one point in history and that was the only time in history, we were just as complex as one other species simply because that species doesn't die at half-life like the earth does not stop turning at dusk. Of course we appreciate that the earth does not stop turning at the end of the longest day.

A invention that is useless has purpose only for the artist but for everyone else as well. A evergreen and the uselessness of a evergreen does stem specifically from the fact that it doesn't bear a fruit we can eat. We all agree we need a evergreen like we need all of the trees, for air, for our lungs, but the only reason we cannot do without a evergreen is because it is the oldest tree in the world. Cooperative labor produces species that lack descendants as uselessness is not considered a positive attribute. Today not even the practices of a ancient religion, which is the simple grouping of species, can we look to, to put order to a populous that was destined to failure at the moment they decided that work and pleasure was not equal and that the day and night is opposite on the political spectrum, that winter and summer are the defining moments of a calendar and are like life and death, opposites. And so the night time became the greatest taboo we have ever known, because animals we don't even know celebrate during the night and species of trees who's life cycles we know very little about experience night time as a season. Fecundity became for the elitists. The simple grouping of species remained a culture kept hidden by the conversations we had during the dark hours of the day.

For cooperative labour to be successful according to Fourier, the passions must be in harmony, so if we gave the example of fruit pickers and tenders, they must each enjoy not only their work, but the fruit they work for. They must enjoy to eat the fruit that they tend to like a winemaker enjoys his grapes. But within the common practices of cooperative labour, the problem may very well arise that a fruit tender may not enjoy taking care of a certain tree as much as another because he does not enjoy the taste of the fruit of that tree as much. And what if no one enjoys that fruit? That is a evergreen. One cannot tend to a evergreen tree because it bears no fruit that we can enjoy eating, the tree only bears seeds that we can use, but it drops it's seeds on it's own and we have no need to pick it's seeds. The evergreen has one attribute which makes us to desire not to be involved in it's reproductive cycle and that is that the evergreen tree produces seeds that will fertilize on the tree, the seed will fertilize on the tree and become a ready seed to become a offspring three years later, and during that time already one calendar year has passed, and two and three: a evergreen tree drops its seeds like we would plant the seeds for a fruit tree, which is a season later, or two or three, even though a angiosperm seed should sprout a year after it is dropped. When cooperative labour is a concern, and the job is to tend to the fruits of trees, most of the fruit of a tree are eaten if not all, of that tree,- say that the fruit tree is a first generation parent tree, so that would mean that it is young, but it is old enough to produce edible fruit that we enjoy. And that is why it is a fruit tree that has become useless because it no longer in most cases produces fruit that we cannot eat, that means it no longer produces fruit with malformation. It is still in the landscape.

The fruit that is not eaten on a fruit tree must have some malformation or disease, but it has seeds, so it will drop it's seeds and if nature nurtures the seed the tree will produce a sapling that will become a tree that bears fruit like it's parent tree. The tree will only be a descendent of it's parent tree because not all of it's fruit was eaten. Let's say that tree is a first-generation parent tree that had one malformed fruit. The malformed fruit, once it grows into a tree of it's own, is a offspring and a tree. But it will take a long amount of years, let us say five, to even be considered a tree. In between those years it's a sapling and a offspring but it is not strong enough to be fruit bearing. Once it is, all the fruit it produces in most cases will still be malformed, because the tree will not be old enough to produce edible fruit, fruit that we find tastes good. In a system of cooperative labor we would have to let the malformed fruits fall, and that is how descendance within fruit bearing tree species has continued in the advent of cooperative labor. The advent of cooperative labor was a one-time event. If all of the fruit from a fruit tree were eaten, and that is a first generation parent tree, and we harvested the seeds and grew new trees from it's seeds, they would be relative trees rather than offspring to the tree, because a tree that is plucked bare during the harvest season will not have it's seeds planted until the next planting season according to our practices, and we want to keep our practices alive. That tree will have it's seeds planted half a year later during spring or a year later during the next harvest season, and this is in a region where the territory goes through four seasons, but in every region seeds are sowed according to season. It is untraditional for us to eat the fruit and plant it's seeds the same season because according to the practices of cooperative labor, it is a invention itself to store seeds and plant seeds and to eat the fruit joyfully without need for consideration to work first before eating. To sow seeds seems a much more profitable labour because, in the case of trees, new trees will grow from this process, trees that can live hundreds if not thousands of years. But is it more profitable of a labour than picking the fruit for ourselves, to eat?

    The next season, those who are appointed the task to plant the seeds picked from a parent tree, will have labored to eat from the same parent trees the clan ate from last year, like the rest of the clan, but other members of the clan will be picking fruits that contain seeds that come from the same parent tree. However last season's seeds that have just been planted are relatives now to their own parent tree. Why the seeds we planted that we took stripping bare that tree last season are only relatives to that tree is because the seeds are a season old when we plant. This creates a cycle of relatives. Even we don't have as many children. This creates a cycle of relatives, some type of informal kin, some type of gathering of species. What the ancient Egyptian god's watched over, their agricultural system, was rooted in tilling the soil meaning they planted seeds the same time they picked the fruits to eat for themselves, their system did not exist around storing seeds, and as a exclamation as to how beautiful it is to use from a flowering tree that grew on its own, we know that good perfume makers never make perfume from a flower or flowering tree that was planted by our hands.

    Any parent tree must produce their offspring in the first generation, when in all cases for a fruit tree the days will sooner be getting smaller and the nights will be getting longer. A parent tree should drop their seeds on their own. The invention to store seeds produces relatives but really almost not ever similar to that type of kinship, in actuality they become sisters. From the advent of cooperative labour and afterwards, angiosperms that take part in it, only really have a first generation, they can produce a second generation, a third, they can produce up until the one hundredth generation, even more, but trees are not supposed to have sister trees, and we do call them sister trees. Naming trees that were to be offspring, that we made relatives, sister trees, makes the strongest species on earth feminine only. Why have we decided to make a sisterhood of trees? Let us debunk even olde methods of tilling the soil and say that they're not good enough.

    Why is night the second day? There are only two days in the generic calendar and two days in the original calendar of life itself. One is the light filled day, and one is the night, the second day. In the chapter genesis, from the bible, it is stated that evening and morning are the first day, and that is true, but if we look at our modern calendar, if we perceive from our modern post, -when even the trees and agricultural society become modern,- the sapling sprouts at night, during the second day. When have we really viewed the sapling sprout? As if it is some sort of magic that brings the sapling out of the ground, when we cannot see it. When in the morning, when we wake up, we see that the sapling has grown out of the soil, and some decided to call that morning the first day, but there was a first day when the seed was planted and whether or not that time was day or night, that was the first day. When tilling the soil became work, that was when we started to disengage with biblical scripture. If we say that happened a long time ago, then we have to revive the old testament, but because we are conservative, we lay the blame on the industrial revolution, when our language changed, when we attempted to arrange the merry-making of a family of trees.

Night is the second day. Though it is stated in the bible that morning and evening are the first day, which they most definitely are, the night time has increasingly become a time when people work and earn money. It is almost sacrilegious but it was inevitable that night would become the second day for us, possibly the final reason for us to call the night our taboo, because we thrive and survive as cities within Canada off of a very profitable electric grid system, it may even be cheaper to employ people during the night than the day because the lights turn on our streets at night automatically, and this has been happening since we created the electrical grid. In Canada, we have never paid for electricity and if anyone complains that they have, they are equating themselves to the indignant man, and there can only be one indignant man, and he is never in a religious mood, he's a farmer that bears a striking resemblance to the one in the painting American Gothic painted in 1930 by Grant Wood. The day time is particularly expensive in Canada because the streets are wide, the distances are far to go to work and in the winter by the time the snow is cleared it is night. The night is also romantic, too romantic in Canada. If night was the first day like god has said, we would never want daylight to arrive again and we would kill for the night and the comedy would become a tragedy: we should honor the people who can survive working through the day, they have kept a comedy alive but those who work during the night are so privileged that they are permitted to be rebellious to the law of biblical scripture, so much so that they have the privilege to call the night the second day. Even the trees agree on some things, as they hum to the vibrancy of the moonlit night.

We plant seeds for angiosperms in most cases in spring, about three months earlier than when the fruits are ripest but a first generation of offspring for a fruit tree can only be a first generation if it was planted when the seed has just been plucked, because that is usually when the seed will drop on it's own, and so that is during fall. The sexual activity of a tree, until it drops it's seed, initiates a relationship to solar and lunar light in the seed which ensures that a tree may inherit progeny that will grow far distances away from the tree. The child of the tree must still not forget his parent tree if other methods of seed dispersal are employed in the future for his offspring. Especially in a world where a seed may not inherit this, -this distance from the parent tree-, it seems important that night makes it's mark on a moment of life which will definitely grow, like a seed. A animal may follow the same path to plant a seed from a parent tree every year, or every season. That pathway is inherited by the parent tree and that inherited pathway competes with the electrical wires underneath the soil. So, for the kingdom Plantea, in Canada, where the electrical grid is perfect, the trees compete with a very old system of natural invention, so that they can grow. In the newly developed communities, in these neighbourhoods, we have planted trees. It is a optimistic estimate that these trees will grow old as well, surely some of them have inherited pathways for their seeds to go far too. All angiosperms in their seed phase are more receptive to lunar light than solar light. The moon at night harbours a pattern of growth, the light from the sun grows larger and smaller as the shadow of the earth engulfs the light and lets it grow, and this light is a reflection of the sun. Whether the shadow is growing bigger or smaller, during the night, the moon *communicates* what some would say is due to a reflection, instead of simply shining, however it should not be uncommon to believe that the moon does hold a light of its own, sometime and somewhere in-between this pattern.

"And god said let there be light: and there was light. And god saw the light, that it was good: and god divided the light from the darkness. And god called the light day, and the darkness he called night. And the evening and the morning were the first day". All seeds are more receptive to lunar light because that shadow and light, that momentus motion, the dieing of a sun on the face of a moon ensures there will be some type of food, some type of sustenance for a seed, for a tree, for a family of trees without worry that that reflection will one day die, it is a contingency that we cannot have for the sun. A contingency we can only recognize to have a existence, if we are awake at night: a relationship to light and shadow, and then light. All seeds are more receptive to lunar light now, but even during the time of the ancient Egyptian gods has the world been this way.

During the life-cycle of a seed, at some time, there are always two elements, two competing elements which is what is male and what is female: male and female gametes. So a seed of a fruit tree for example that will sexually fertilize for a year will relate to night and day as two spans of time that are male and female because, what is inside that seed will have to introduce itself to the celestial environment and be either male or female, and proclaim that he too knows of what the intrinsic law is, which is that there will always be a male and female and there will always be a man and woman. For the seed to sprout out of the soil, the seed targets the maternal womb of a night time that does not have a maternal instinct. What is night does not become female, but the ever changing pattern on the face of the moon resonates a rhythm which is the same rhythm that resonates to the hips of a woman and her menstrual cycle and to the anticipation of progeny for a woman and a man. Upon the sexual fertilization of the seed of a flower, the growth of the seed experiences sexuality in relationship to the celestial environment. Because sexual fertilization does not happen quick enough for the flower, to notice a night sky and to notice the weather becomes probable. The seed of a flower cannot resist the celestial environment it is introduced to during sexual fertilization, this is why pollen spreads through the air, and can solemnly detangle absurd notions of misplaced sexual desires during the day, remember, pollen is the male gametes of a flower. *The seed becomes a existent, but why it is prone to admire the light from the moon is because the second day will most likely be the time when the plant will sprout out of the soil, the second day being the night:* says a silver fox tree. And the night is female to the seed, not because of any maternal darkness we relate to a female womb, but because the moon that harbours the reflection of the sun is male only if we live within a comedy, and a tragedy, forever, only if we live in a world where natural opposites are detected in the face of the unfortunate arrival of inaccurate myth. And isn't it tragic that the chance of a fruit tree in a world of cooperative labour will only produce a offspring if it could produce just one diseased fruit? If we planted the seed of a angiosperm in august when it would most likely be dropped, it would begin to grow in august when the days are getting shorter, and it would endure winter, for five or so years be a sapling, until it can grow its own first fruit. But the moon is what harbours the weight of the sapling, because only the moon can provide a reflection of light unto our earth and our plant species. That reflection of light causes our first generation of plants to reflect, to grow in the image of their parent tree because the patterns of the shadow of the earth that cross the moon tell a sapling that: *though the sun will one day die, I count away the hours and days and years and decades and centuries until it will die, just like though you will one day die, I count away the hours and days and years and decades and centuries that you will one day die, so that you will choose to grow and look like your parent, because you do not have two parents, and nor does the sun.*

Because at one point in Greece on their monetary coins was the image of a grand tree. It's too intricate of a image to see if that coin does in fact have a emblem of a fruit tree engraved on it like the ones we've seen, but I can remember a tree that is large and wide as a big flowering tree would be, and I think that coin was in circulation during the era of the Greek gods. Cooperative labour cooperates according to the sun, we plant as the days grow longer. This takes any monetary value out of our labour because we never eat just half a fruit to plant it's seed, or a bit and throw it to the ground to plant a seed, because that would be a waste of tax-payers money, but we are the tax payers, and we prefer to eat the fruits of our labour instead of enjoy the fruits of our labour because, to be sincere, do we really want to eat every fruit off that tree, and make use of it best we can, or would we rather eat when we want, what we want and waste nothing of a useless tree.

Chapter 4

Never Fall Prey to Heliotropism

Van gogh and his political agenda to capture heliotropism in a painting proclaimed through directional brushstroke that light travels rather daintily and that electrical light and sunlight and starlight and moonlight travel in the same direction, a direction which resorts to the idea that all things revolve around the sun. We may include only two of the same species in our biological classification system that we paint, and these species are not sexual opposites or of opposite gender, it is yet to be seen when the opposite gender has a taxonomic identity, when we include a she in the chart of life from the animal kingdom. If our biological classification system that we paint curiously includes two stars, one the sun of course, and one, what is a star, then it will be made certain that our sun is part of our celestial sphere because it is a star, but our moon is not part of the celestial sphere because it is not a star. These two species, one what is a star and one the sun, they are the same species and they curiously exist on different parts of one lifespan, but of course the only reason as to why the sun exists in our classifications is because he is named! The moon is also named and he was once a star, but because he is not one anymore and so he no longer exists in any measure of magnitude, he remains a solumn moon. What Van Gogh did in starry night, painted in 1889, was uncover a horribly hidden secret that the brightest star in our skies can never be seen at night, the sun can never be seen at night, and our perilous view of the celestial sphere put us in a position to refrain from celebrating during the day because we cannot celebrate the procession of constellations with just one star. Only one star out of all of the many we can see do we see during the day and we cannot celebrate one star, -like we can celebrate the north star, as it is, alone,- for then we would be celebrating the earth's very own revolution around that one star. We would be choosing to celebrate one star as compared to a innumerable amount of stars. We have to make the choice of as to when we should celebrate because our political views are at stake if we don't make that choice. Choosing to celebrate at night makes our night time our faction, our impulse, our depository for every piece of artwork until the death of modern art. All, most all of our paintings were done during a day-light filled day. And during nights indoors with the light of electricity. Or outdoor at night with the help of a light that could be electricity, or could be a oil lamp or a burning fire. But if the painting were done under the night sky with only the light of the moon and the stars there is no way to prevent that painting to become of the celestial sphere because if you don't paint the sky at a moment like this, you will end up painting a myth that cannot be veiled: the depository of all our paintings since the death of modern art is the night, the act would be like destroying a taboo. We don't always know at what time of day the painter painted. This aspect of fine art makes very few pieces of art conservative in nature when seen in a natural setting and when seen in a environment that is not lit by electricity. The true definition of hedonism is restraint in the face of love, a unending condition of control every painter fights with in order to make a piece of art. Each painter fights with a myth, keeping hidden the myth, revealing the myth, there is nothing as true as the control of the paintbrush, there is no aestheticism what so ever in art and there has never been. What causes us to paint fine brushstrokes or messy ones is our desire to keep hidden but reveal just enough of a myth. It is the myth that is beautiful in a painting. Once you discover what it is that is existing between a work of art, you will find that work of art more attractive because then you know you might be closer to discovering a treasure, because every work of art until the death of modern art is just a route to hidden treasure. If we are conservative in nature because we celebrate the night, we are no conservatives without the day. However, if we view the paintings we painted during the day at night when under electricity we can become vain just so we can celebrate the procession of constellations without one star.

We at some time during history made a globe of our celestial sphere out of plastic. It was light to carry. We probably also made one out of wood. But in making the globe of the celestial sphere from the viewpoint of outside of the sphere, -from the view of a onlooker to a bunch of stars,- one too many times, we did nothing but create a plaything, a interesting object in passing, because the zoo that is our constellations was already locked up and tied up into the box which was that earlier invention. That invention is the very first globe that mapped the land of our earth. The constellations could never have become a celestial sphere, one that we could hold in our hands, because long ago we made up their regions and placed them as hemispheres on the globe, and, because the celestial sphere is hollow! The northern and southern hemispheres of the constellations and of the globe of the earth, the northern and southern latitudes in relation to each other are the firmament above the firmament and the firmament below the firmament. This is why east and west, and east to west only exist on a map, because we painted the first globe with the fingerprints of the constellations that surround us, and there is no east and west in that world. We must have made that globe at night, and so there the zoo of our constellations were lost to our territorial globe and the age it represents. There is a east and west on our earth in many ways, political, religious, and that is a significant mystery and it is that socio-political and religious east and west that originally gave the globe age, certainly from the days of the fertile crescent on, when Oceania was also a boundary around the center of the mediterranean sea. When only the Egyptian goddess Nut was the sky above that part of the sea. When the fertile crescent was just that.

We began to make borders and boundaries but we lost the zoo to the globe of the earth because the four hemispheres on a paper map of the world did not correlate to the four seasons. The best way to believe is that the four seasons do not exist because of the earth's tilted seat in space, or the earth revolving around the sun!

As if the earth and outer space create some type of machinery that every star in our celestial sphere is hinged to, to ensure we have seasons.

The paintings we have from the renaissance of a super round earth, a earth that tilts on no axis, are better paintings to reference when we are looking for answers about what our earth looks like in space, and why we have four seasons. If you want to believe, sometimes, that the earth does tilt on a axis, because perhaps you do like the sunset, because that tilt is romantic, still, even Monet's sunsets may not adhere to your taste, you may be better suited to work in the romantic perfume industry!

We lost the zoo which is our constellations to the globe of the earth, and it will stay lost there, because those old paintings from the renaissance and before, of a round earth, a super round earth, make the earth so enticing, that anyone who studied art history would make a map of the earth now. The artist would make a map of the earth with the celestial sphere painted on it, from the outside view. And would prefer to make it a globe, that spins from right to left, and would make sure that it is not spinning on a axis. On a globe, unlike on a map of the earth, there are only two hemispheres, the northern and southern, like the northern and southern latitudes of the celestial sphere.

But when we created a globe that mapped the earth we imagined ourselves atop the earth.

What we did when we animated the celestial sphere, when we put a map of the star constellations on a globe where the earth was already mapped, when we stood outside of the celestial sphere, we maintained the myths surrounding our star constellations and singular stars and star clusters that make up those constellations, and we did all of this while already existing in the heavens, here in Canada. To make a globe like that, and to make a map of the earth, from our land, from where we already exist, from the heavens, is special. *That was a zoo we could relate to our calendar and our seasons.* And when we do make a map of the earth from our unique viewpoint, as if we are standing and facing the universe in awe, on Canadian soil, a flat map becomes quite precious to us too.

The maps we make in Canada are precious. It could be because of the paper we use to make these maps, because certainly that will be good paper. It could be the gold and black ink we use, because that will be good gold and black ink, but also it is because this map will be made in heaven.

The globe of the constellations from the outside view could have been a work of art but we used it greedily for work. We have made many globes now that animate the constellations which make us onlookers and insiders, most of them are not works of art, they are inventions we can't Really use, a useless invention.

A globe that maps the constellations is a work of art if it is a original and lone. If it were the unimaginable and unfathomable globe of land and constellations that makes us a insider to vast space, it must already be in a museum.

But, only one globe full of characters of our night sky could have made a inventor very rich, a artist well appreciated, a collector charmed, and that would be a globe of the night sky alone, from the view of the outsider to the celestial sphere, that spins towards the left, because then and only then would we have a machine that works like a machine that revolves should work, and then the night sky would not be partial to a prevailing mystery, as to why the direction of the movement of the arms of a clock regulate and rule that every living machine, like heliotropism, must turn to the left, and disbands only upon the political spectrum.

Summer and winter solstices are conservative celebrations and to be part of them you almost have to be a angel, or well-read, and a recluse and what is more angelic. During these celebrations we revel in political disputes more than any other topic. Whether the longest day should be called the shortest night and the shortest day should be called the longest night is not small talk at these parties. We do talk about astronomical events on these days, and we talk about Orion's belt and the moon. If we are celebrating summer solstice we are also talking about a cup full of wine and picking grapes that are ripe off the vine. Despite what is popular, Orion's belt is the favored part of the entire constellation of Orion for us. Especially on a summer solstice it is important to discuss how a woman can be a leader in our world, we already have a celestial sphere that is led by a woman, a woman with great birthing hips. Summer solstice is also the best time to talk about heliotropism.

Because in our biological classification system we have two stars, one the sun and one what is a star, by not giving what is a star a name in our chart of life, we will only incorporate the life cycle of one star which is our sun in our classifications, but of course we know the one star that is not named but is in this chart has a life cycle, it represents the life of more than many living stars. We may know the life cycles of stars of different magnitudes. We know of their nebulae beginnings and black hole or supernova ends but stars are not like the life we have on earth, where each tree or flower pollinates and animal life mate. The life cycle of a star is seasonal unlike any life we have on earth. A star has it's own season, represented by colours, bright shining colours. That is it's life cycle, and that is why, though we believe stars are animate like we believe the sun and moon are animate, and though we have a very close relationship to the constellations that serve in our night sky, stars are left for myth alone, and they are solumn pieces of land to us. Stars, including our sun, have lifespans but they do not live in a weary resignation to a half-life, and this is because our sun has passed it's half-life, and this happened right before our blind eyes as the gears and levers began to move towards a more liberal world, when we made woman equal to the sight of a blind man. When we tempted fate and called ourselves subversive. When women began to act on the silver screen and play in rock bands. When women almost equaled wages with men in the working world, but, it was never expected that a woman would ask for a muse.

This muse that the woman has asked for is not the medium of one art form and is not a husband. When a woman became a popular artist in the world the five art forms became a womb because a woman discovered for herself that she did want to have children, despite the fact that she could defend herself and her individual identity as one of the artists. And when she became a artist, a woman was not wronged by the age of convenience, because art work was still hard to create when she did make art. In the modern world, today, this woman, this artist thought that if she gave birth to a daughter that her daughter would pass down a tradition from the mothers before her, which is not to be confused in the face of diversity. A woman thought that she would have a daughter and her daughter would also be confident when she was called upon to conquer the five art forms. So even if this woman cannot know if giving birth to a daughter is a act of creation, because she is a creator, at least she knows one thing about her daughter when she takes a prophetic glance into the future, something she does not know about her son, which is that a daughter born from her will conquer the five art forms.

And she does know that her daughter will be good with money, because in the face of diversity, when the five art forms will try to rattle her cage, she will be good at all of them, because we live on a land where the resources to make good art are available. A young girl will become a great woman and she will make something of herself here in Canada, and there is a big business here which will make sure she will and it is not different from Canadiana: in the modern world, here in Canada, a woman will still weave, and sow, and quilt, and paint and that will always be considered art.

As for the desire to give and the desire to create that a woman lives through: In the modern world there is still no confusion for that woman if she should do any of this or do all of this. But for a woman that gives birth, for a womb that is predestined to give birth to a boy and a girl, the act of giving birth to the two sexes is different. It seems obvious to me that the desire to give birth to a boy is a act of creation, and the act of giving birth to a girl is some other act, and then perhaps we could begin to believe that the ovum does attach itself to a specific place,- with the help of the momentum of a spermatozoon,- in the uterus of a woman, according to what gender will be born. As if, if we viewed with a microscopic lens, that there is more machinery to the uterus than we have already found.

There is nothing wrong with normal. There is nothing wrong with a very long tradition of habitual patterns that the body of a woman propounds as important, which could simply mean that to have more than one child with the same man is made easy by biology itself. Perhaps the spermatozoon makes a map of the uterus it enters, and so that man who is part of the reproductive act with you, is a man you should stay with. The animal cell, that took part in the sexual act with him and you, could even promise a sibling, any day of the week! Even the first child that is born to a woman and man, whether it was a planned pregnancy or a pleasant surprise, that baby was born to a man and woman and from the body of a woman, as a result of habit...*that the body of a woman and the mind of a woman is always prepared for a pregnancy.*

Since the birth of a baby girl, that baby girl has monopolized sales in Canada because when that baby girl gets older she will need menstrual pads. Menstrual pads are the most sold item in our world. So a event in the life of a young girl, who becomes a woman, has now become a event which is celebrated in our world, because when a young girl or a woman wants to spend money but she has everything already, -she has nice clothing, she has jewelry,- she has all of the fancy items in life, she can find a excuse to spend money on a fancy item that she will always need and that is the menstrual pad. Even when a woman goes out into nature, would she not want a menstrual pad? And so quickly, in our beautiful world, a event in the life of a young girl and a woman, which is menstruation, has become a creature comfort, because a young girl can enjoy shopping for a habitual pattern that happens to her body, which she cannot avoid and our world in the west does not want to avoid it either.

The young girl has monopolized on a normative society, and on the generic calendar.

In the world outside of our home a creature comfort is a purchase that we enjoy in the normative world, like buying another coffee from the place you go to often, or perhaps from somewhere new, the experiences that overload us with plentitude, and this habit is political. Despite the reality that buying a coffee from the place we often go to is fun, on a very tough day, when the sky is cloudy and it is raining here in Canada, or it is snowing and it is cold outside here in the neighbourhood, to enjoy a creature comfort is a decision to not be bored. This habit is also a profound statement about the world and about the way people think: We are not afraid to have fun, to find comfort, in a beautiful world. But a young girl has monopolized sales: the first creature comfort to ever exist for a woman, and her household and this world, was the arrival of a young girl's period. And it is the first creature comfort and the only one that has created a pattern amongst women, according to the beginning of life, this pattern to rear a child, a young girl, up until the age she receives her period, and from then on she is expected to take care of herself? Medically speaking, in our normative world, a young girl becomes a woman when she gains her period, because most of the investigation of the female body has been based around the adult female. But the desire to give birth to a baby girl from the feelings of a mother is a desire to give birth to a baby who she can educate and protect and defend, because the body of that baby will one day be like herself. A mother wants to be the only one who can investigate the body of her baby girl because there has never been a map of her body and there never will be. When in infancy, the baby girl is alone with her mother, there is no boy around to make a map of her body. And this is where the desire for a woman to have a muse came about: she will put her in dresses. A young girl is born out of a desire, whether it is a act of creation or not it is still yet to be found.

The End

www.ingramcontent.com/pod-product-compliance
Lightning Source LLC
Chambersburg PA
CBHW031415210526
45464CB00005B/1900